KING OF THE COMSTOCK SOUTH

THE LIFE, LETTERS, AND LEGACY OF LEWIS CHALMERS

KAREN DUSTMAN

AMERICA
THROUGH TIME®
ADDING COLOR TO AMERICAN HISTORY

For all who love the magic of undiscovered history.

Cover photo credit: Jay Topping

America Through Time is an imprint of Fonthill Media LLC
www.through-time.com
office@through-time.com

Published by Arcadia Publishing by arrangement with Fonthill Media LLC
For all general information, please contact Arcadia Publishing:
Telephone: 843-853-2070
Fax: 843-853-0044
E-mail: sales@arcadiapublishing.com
For customer service and orders:
Toll-Free 1-888-313-2665

www.arcadiapublishing.com

First published 2023

ISBN 978-1-63499-445-3

Typeset in 10pt on 13pt Sabon
Printed and bound in England

Acknowledgments

Grateful thanks to Jay Topping, Judy Wickwire, Ed Rogers, the Alpine County Historical Society, the California State Library, the Society of California Pioneers, the Beinecke Rare Book and Manuscript Library, and the Huntington Library for kind permission to share images in this book. Special thanks also Debbe Nye for research assistance and to Carol A. Jensen, whose initial encouragement brought this book to life.

Several of Lewis Chalmers's original letterbooks have not yet been found. My hope is that this book may prompt further information about this amazing man to come to light. If you have additional data or research materials to share, please contact the author at *KDustman@Clairitage.com*.

Contents

Introduction

Our phone rang in March 2005. The curator of the Alpine County Museum needed someone to drive over the mountains to Ceres, California, to pick up a donation of old furniture and some letterbooks. Could we help?

The trip would require a 150-mile journey to California's Central Valley, and honestly, the furniture he described did not sound all that exciting. But the proffered donation was by a member of the Whitmore family, former owners of a local landmark known as Chalmers's Mansion, and the brief mention of letterbooks piqued our interest. These supposedly contained correspondence by the early mining capitalist Lewis Chalmers himself—how could we resist?

A three-hour drive later, we pulled up in front of a stately Victorian home that once could have rivaled San Francisco's painted ladies. Now, however, its paint was peeling and the yard had gone to weeds. Financial woes had overtaken the owner, he explained, and he needed to move soon. But before he left, he wanted to make sure that bits and pieces of Chalmers memorabilia found their way home again to Alpine.

There was a wooden table in rough condition but reputedly crafted by Chalmers's own hands, a couple of unimpressive ladderback chairs, and a clunky wooden table base, minus its top. Nothing that would make a museum curator terribly excited. My heart fell. Had we made this long trip for nothing?

But then the donor pulled out three large, battered cardboard boxes, stuffed to overflowing with leather- and cloth-covered volumes. The boxes smelled as if they had spent the last century in a damp basement. They probably had, but oh, the books inside. Letterbooks and ledgers compiled by Chalmers himself—decades-worth of them.

The ledgers recorded the financial minutiae of Chalmers's early mining ventures in Alpine County, and the letterbooks, our host explained, were how correspondence was preserved in pre-Xerox days. Inside, the tissue-thin leaves were semi-transparent. So a writer would first pen his letter in water-based ink, then wet a page of the copybook and place his letter beneath it. The two would be firmly pressed together in a press, transferring ink from the letter to the backside of the tissue paper. *Voilà*—a perfect duplicate of the original letter was preserved in the bound letterbook.

On our drive back over the mountains, my husband and I wondered what information could be hidden in those pages. Little did we know, at the time, just how deeply we had become involved with their contents ourselves.

In the years to come, I transcribed most of those letters—a slow, laborious undertaking involving thousands of typed pages, but in the process, I came to learn so much about Alpine's early mining days and about Chalmers himself. What an astonishing record he had left us.

Surprisingly, as prominent as Lewis Chalmers was back then, he is a virtually unknown figure today. The only trace of his existence most people recognize is a modest house on a California back road outside Markleeville, still known to locals as Chalmers's Mansion.

There is more than a touch of irony there; the house is hardly a "mansion." Yet from 1868 to 1885, Chalmers truly was king of the Alpine mining region. He touted its mines as a "Second Comstock" and promoted his Isabelle Tunnel as a "Second Sutro." For almost twenty years, Chalmers invested both his own hopes and a flood of British money into a flurry of ambitious mining ventures.

This is the long-forgotten story of Lewis Chalmers and the untold legacy he left us in his letters.

1

The Fraserburgh Years

Lewis Chalmers was not exactly born to royalty, but for Fraserburgh, a small fishing village tucked away in the far northeast corner of Scotland, he was the next best thing.

His father, the elder Lewis Chalmers, was Fraserburgh's chief magistrate and baron bailie.[1] His mother, Jessie, was the eldest daughter of William Kelman, who himself had been a bailie. So when young Lewis entered the world on March 8, 1825, exactly a year and a day after his parents' marriage, he arrived with a proverbial silver spoon in his mouth.

The first of what soon would become eight children, Lewis was showered with all the benefits wealth and social connections could provide. He was sent to the best tutors and finest schools, moving on to study at Marischal College at Aberdeen. There he excelled, winning prizes in Greek and Latin. In a sign of his early fluency with words, Lewis won a third prize in arithmetic in 1840 but took first prize at writing.[2]

After his graduation from Marischal College in 1843, Lewis dutifully followed in his father's footsteps, burying his nose in law books at Aberdeen and Edinburgh. He was admitted as an advocate (lawyer) in Aberdeen on June 29, 1847, at the age of twenty-two.[3] That same year, Lewis became an agent for the Scottish Provincial Assurance Company and secured employment (as had his father) as an agent for the Bank of Scotland at Fraserburgh in 1848.

Already a polished gentleman, his career trajectory seemed set. "He has always moved in the best society," a supporter would later enthuse, "and has the manners and address of one who has done so."[4] But young Lewis was not without a wild streak. In April 1850, he was forced to acknowledge an illegitimate child he'd fathered with Elizabeth Moir of nearby Banff. The baby received not only the Chalmers surname but also a clever variant of her father's first name, as well: Lewisa.[5]

The illegitimate birth must have caused quite a scandal in the neighboring small communities of Fraserburgh and Banff, and it is possible young Lewis received a stern parental admonition to settle down. If so, it seems he took it to heart.

On June 25, 1850, less than two months after the baby's out-of-wedlock birth, Chalmers married Eliza Ann Gordon Cameron. Eliza was the eldest daughter of

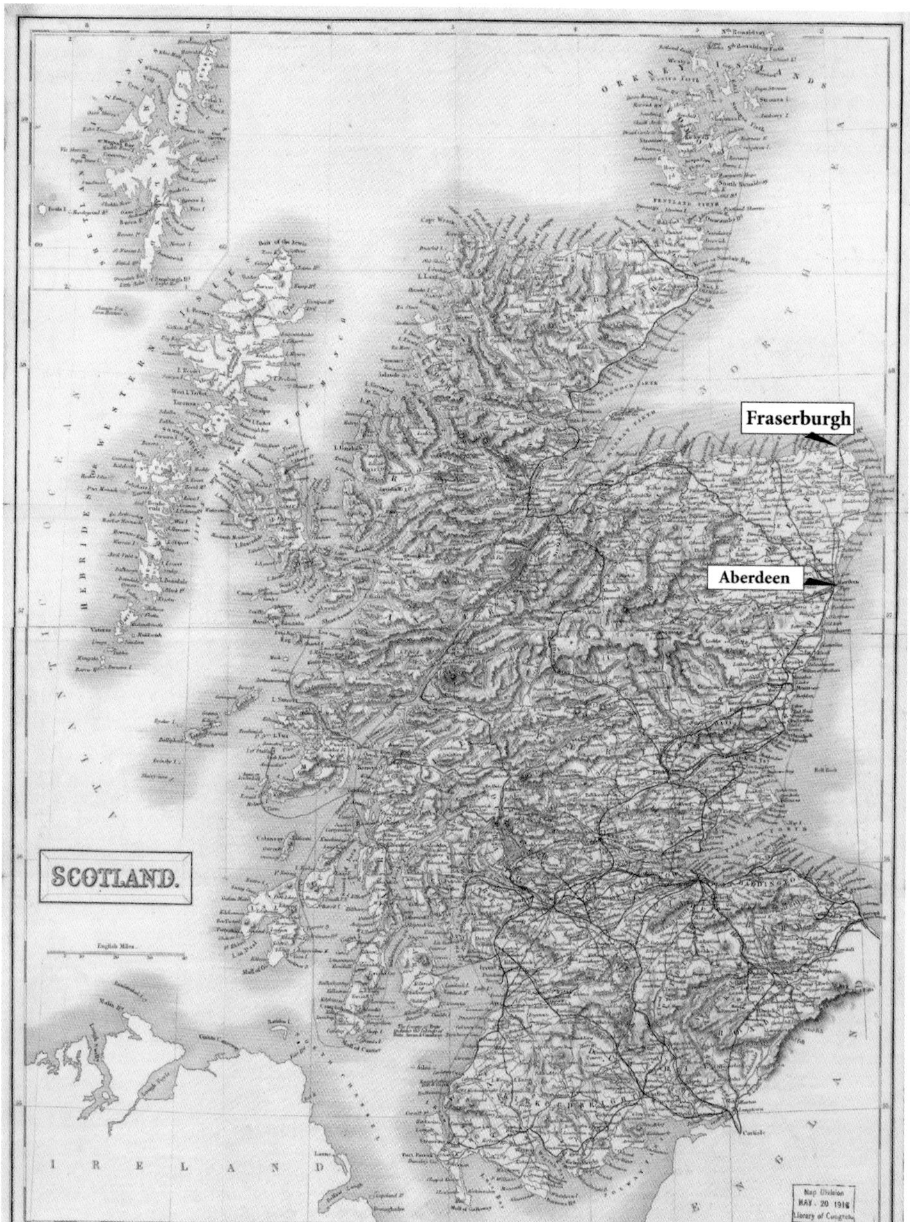

Map of Scotland, *circa* 1855, annotated to show Fraserburgh's location in northeastern Scotland, beside the North Sea. (*Library of Congress*)

wealthy James Cameron, and their marriage was celebrated at her father's lavish seventeenth-century country home, Aberdour House.[6] The couple would welcome their first child, a boy, a year and a half later, on November 30, 1851. This child's name, perhaps not surprisingly, was Lewis.

Despite such auspicious beginnings, twin tragedies soon descended on Chalmers. His father had passed away on September 20, 1850, at the age of sixty-four. Slightly more than a year later, on December 6, 1851, wife Eliza died, too, just a week after baby Lewie's birth, from a childbirth-related infection. She was just twenty years old.[7]

Lewis Chalmers's father had left an immense estate of over £6,000 (approximately $925,000 in today's dollars). That translated into roughly £750 ($116,000) each for Lewis and his seven surviving siblings. As part of his share, Lewis had been asked to take the family's twenty-four-room home in Broad Street in downtown Fraserburgh and to provide a home there for his maternal grandmother, Jean Kelman.[8] This he did.

The 1851 census found Lewis, then age twenty-six, "head" of the household at the Broad Street address. His extended household included not only his grandmother, "Jane" Kelman (age seventy-eight), but also three sisters (ages seventeen to twenty-three); a cousin; wife Eliza (just before her death); and seven servants. Chalmers gave his occupation as "Bank of Fraserburgh/J.P.," and made sure a side note was added to the census to reflect that he also was a notary public and agent for the Bank of Scotland.

Eliza's untimely passing left Chalmers with both an infant to raise and an elderly grandmother to care for, and he was understandably eager to marry again. He ventured down the matrimonial aisle for a second time on July 5, 1853, wedding eighteen-year-old Ellen Miller MacEwen at Brighton, a daughter of David MacEwen. By now Chalmers was twenty-eight, a decade older than his new bride. They wasted no time beginning their own family, and their first child, Alexander, arrived almost exactly nine months later, on April 28, 1854.

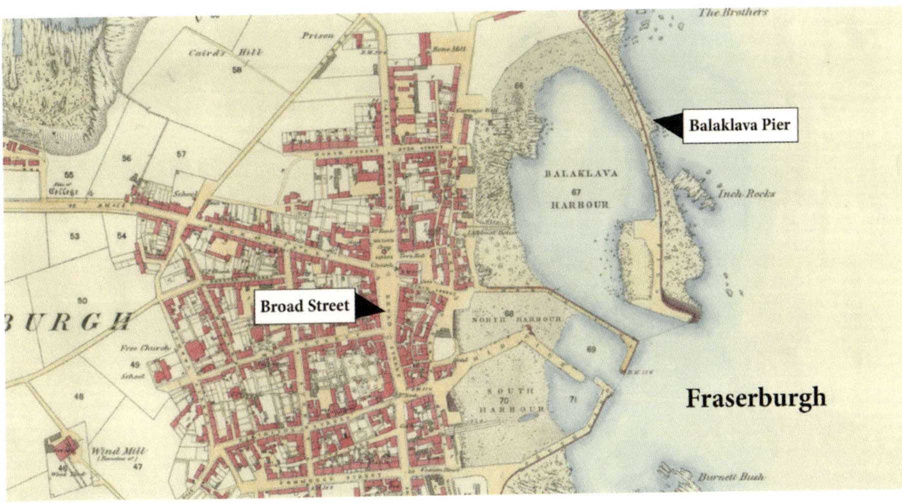

Map of Fraserburgh, Aberdeenshire, showing Broad Street's location in the heart of town. (*Creative Commons; reproduced by permission of the National Library of Scotland, Map 74479967*)

Lewis had long assisted his father in his business endeavors, so he managed to step smoothly into his father's professional roles. Like both his father and grandfather before him, Lewis became baron bailie and chief magistrate for Fraserburgh. He also assumed his father's post as a factor (business agent) for Lord Saltoun, a regional powerhouse. Less officially but just as important, he inherited his father's network of prestigious business connections.

For more than a dozen years, young Lewis Chalmers figured large in the public life of Fraserburgh. He earned a reputation as straightforward and honourable in his business dealings, and he excelled at public oratory. He also won acclaim for his talent at settling disputes, exhibiting both kindness and generosity. When called upon to resolve tenant disputes for Lord Saltoun, for example, he was praised for never "ejecting the widow or unfortunate, but in many cases assisting them with his own personal means."[9]

As chairman of both the town council and the harbor commissioners, he worked hard to bring a railroad to Fraserburgh. He also helped finish several large projects begun under his father's stewardship, including the Balaclava pier, a massive breakwater designed to improve the local harbor for steamships. As if those professional duties were not enough, Chalmers lent his considerable energies to volunteer activities as well, helping form the 5th Aberdeenshire Artillery Volunteers and becoming the group's first captain in 1860.[10]

Chalmers's Broad Street home was situated in the very heart of downtown Fraserburgh, and old-timers would later describe his office as the "chief business centre of the town."[11] He dressed and acted the part; fellow citizens called him "a man of most refined appearance, of great ability, and one of the most accomplished and public speakers that ever lived within the bounds of the burgh."[12]

Chalmers's legal acumen, too, quickly earned him respect, and fellow Fraserburghians attested that as chief magistrate, he "discharged his duties with faithfulness and clearness of judgment." He "intuitively arrived at the conception of what was desirable and beneficial," said one supporter, "and contrived, by the persuasive powers with which he is so eminently endowed, to soothe opposition and win respect."[13] Ah, persuasive powers. It was a gift that would carry him through the rest of his life.

Chalmers's family continued to expand. The 1861 Census shows him still at his spacious Broad Street home, now with four children under the age of seven plus a three-year-old niece. Oldest son Lewie, now nearly ten, was not mentioned and was probably off at boarding school. Seven servants rounded out the busy Chalmers household. These included a nurse and an under-nurse, a cook, laundress, parlour maid, housemaid, and a groom.

Fraserburgh itself was still a small town, with a population of just 3,472 souls in 1861, but thanks to its felicitous position by the sea and a booming trade in herring, the port was growing. Chalmers's own ambitions, it seems, were growing, too.

Sometime about 1862, Chalmers built an elaborate country home for himself on Lord Saltoun's estate outside of town, dubbing it Witch Hill House. When finished, this became a "splendid establishment of hunting horses and hounds, carriages etc., still spoken of with bated breath by the old surviving inhabitants who remember the quite 'imperial' style of living adopted by the bailie," as the *Fraserburgh Herald* later recalled.[14]

Right: Lewis Chalmers smiles charmingly for the camera, *circa* 1875. (*Alpine County Historical Society, Arnot photo collection*)

Below: Map of the greater Fraserburgh area, showing proximity of Witch Hill to the town. (*Creative Commons; excerpted from Ordnance Map of Scotland circa 1868–70, Sheet 97; reproduced by permission of the National Library of Scotland, Map 216386868*)

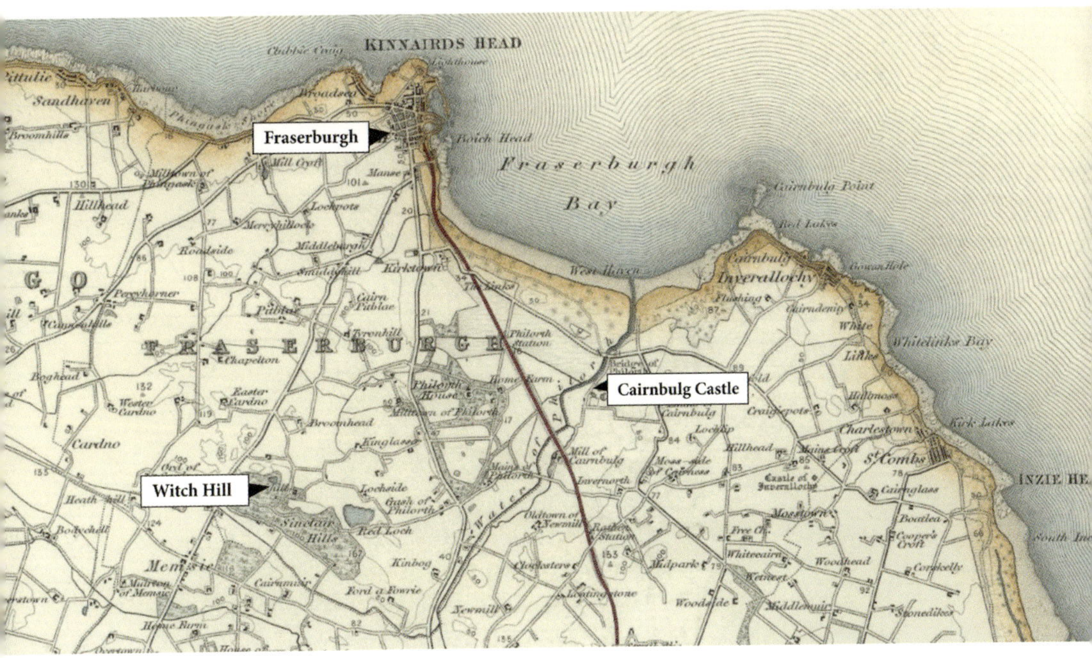

Sadly, it was there at Witch Hill that his second wife Ellen passed away on 23 July 1863, at the age of twenty-eight. Her death was attributed to "phthisis, upwards of seven months"—or as we know it today, tuberculosis. Left behind were step-son Lewie and six of the seven children she'd borne to Lewis during their ten years of marriage: Alexander (born in 1854); Willie (born in 1856); Laura (born in 1857); Jessie (born in 1860); and twins Elleny and Johnny (born in 1861). Another child named Ellen, born in 1855, had died in 1856.

As if this fresh loss was not tragedy enough, Chalmers's financial world came crashing down around him about this same time. Two different explanations have been given for this disaster. The first, as Sir A. Anderson, lord provost of Aberdeen, politely put it, was that Chalmers had been "victimised by a brother-in-law to the extent of above £5,000," apparently thanks to some unwise bit of speculation.[15] If true, it was a staggering loss indeed, equal to roughly $770,000 in today's dollars.

A later historian has advanced a different version of events, however, asserting that Chalmers built Witch Hill House without Lord Saltoun's knowledge or permission. Although Saltoun himself lived not far from Witch Hill at Cairnbulg Castle, he had reportedly been "away for a time" while the actual building went on. In this version, Saltoun became angry when he returned and discovered the lavish project, summarily firing Chalmers. "So it was not finances that caused [Chalmers] to leave but the loss of his job and his respect in the town," according to this explanation.[16]

It is possible that both accounts contained a grain of truth. Chalmers could well have been overly optimistic in his personal investments, and he may also have inadvertently stepped over the line in his management of Saltoun's estate. Given the long history entwining the families, perhaps Saltoun did not actually fire him. But even if they only exchanged harsh words, the proud Chalmers may well have chosen to resign.

Whatever the source of his troubles, Chalmers was now financially embarrassed and made up his mind to leave town.[17] He resigned his post as the chief magistrate and gave up his seat on various commissions, including relinquishing his post as captain of the Aberdeenshire Artillery Volunteer Corps—the force he had been instrumental in founding. In November 1866, his membership in the Society of Advocates in Aberdeen also ended.[18]

Now essentially bankrupt, Chalmers transferred all his assets to a trust for the benefit of his creditors.[19] The *Aberdeen Journal* carried a humiliating notice of the sale of Chalmers's household furniture and effects, to be held June 14–15, 1864. Included in the lengthy list of property to be sold were a rosewood pianoforte, a set of elegant silver knives with jasper handles, three telescopes, and one "well-bred and powerful Horse." Bargain-hunters could paw through the former bailie's silver tea and coffee sets, mahogany furniture, alabaster figurines, and "upwards of 1,000" books.[20] Chalmers was also forced to relinquish his fractional interests in two "whaling and seal ships," the barques *Milenka* and *Alexander Henry*.[21]

The large Broad Street home he had inherited was auctioned off in August 1864. This wound up in the hands of the Bank of Scotland, which would later convert it into offices.[22] Chalmers vacated Witch Hill House about the same time; the following month, his residence was given as "Orton Cottage."

He was bowed but not defeated. Listeners would say that as Chalmers left Witch Hill for the last time, he vowed: "I shall redeem my past and come back and claim my position again."[23]

Witch Hill House, *circa* 1950s. Sadly, this historic home burned to the ground in February 2012. (*Courtesy of George Dey, Fraserburgh Heritage Centre*)

For the present, however, Chalmers felt he must leave his hometown. His major regret was that he had not yet seen the breakwater completed.[24] The railroad extension, too, was not quite finished; the first train would not arrive at Fraserburgh until the following year.[25]

Friends in Fraserburgh took up a subscription and arranged to have a pamphlet of "Testimonials" printed for Chalmers, "in appreciation of the public services rendered by him to the town during the time he held office here as Chief Magistrate."[26] These included laudatory words from a speech by Saltoun himself, praising Chalmers's "energy and ability" and acknowledging his former factor's dedication to the "good of this town of Fraserburgh."[27] Without going into details, Saltoun publicly declared that he "deeply regretted the events which had led to" Chalmers's resignation.[28] Still, it was a huge come-down for such a gently-bred man—one who had managed estates for Saltoun and others since he was twenty-one, and served as Fraserburgh's chief magistrate for seventeen long years.

So what was a thirty-nine-year-old widower with seven young children to do? After winding up his business affairs in Fraserburgh as best he could, Chalmers deposited his children in Ealing, a borough of London, under the care of his sister-in-law, Eliza Chalmers. The wife of Chalmers's younger brother William, Eliza was a recent widow, having lost her husband on May 24, 1864. The living arrangement must have seemed like a good solution for both parties. For Chalmers, it assured that his children would be well-cared-for by a family member; for Eliza, it meant a small income.

Now based in London, Chalmers cast his eye around for new possibilities. The 1859 discovery of silver on the Comstock, a continent away, was still fueling feverish interest in mining investments. To Chalmers, it must have smelled like fresh opportunity. He set about improving his marketable skills in that direction.

Likely through the assistance of influential family friends, Chalmers secured a position with the precious metals firm of Johnson, Matthey & Co. in London, where he could study the art of assaying. It did not take him long. By July 1866, the firm pronounced him thoroughly proficient in assaying gold and silver ores.[29] Here, for better or for worse, fate stepped in to change the course of Chalmers's life forever.

On June 11, 1866, California mining entrepreneur James Jones handed Chalmers a pair of ore specimens from his Florence and Manchester ledges to assay. Two days later, Chalmers certified the results of his work. The yields proved nothing short of astonishing: £1,288 in gold and silver per ton for the Florence; and £435 per ton for the Manchester.[30]

These were only the first two lodes owned by the Imperial Silver Quarries Company, Ltd., a newly-launched British company formed to work a property that supposedly embraced twenty-seven parallel ledges in the Monitor mining district of Alpine County, California.

Capt. James Jones and his brother, Frank, both of Allegan, Michigan, had been among the first arrivals in the Monitor mining district when it began to boom in the early 1860s. They had claimed this location in 1863, dubbing it the Michigan Tunnel Co. By 1866, a tunnel it may have been, albeit a short one. Yet little actual mining had been done there, however, and it was a dubious mineral property at best.

Nevertheless, Capt. Jones had traveled to London where he talked up the claim, and the newly-formed British Imperial Silver Quarries Co. had acquired the Jones's mining title on February 23, 1866, hoping to float it for investment.[31]

Despite its glorious name, the Imperial Silver Quarries floundered at first, failing to attract investor enthusiasm. But a year later, in June 1867, just when the Imperial was "almost at its last gasp," a financial angel appeared in the shape of Lord Poulett of Buckingham Gate. Poulett agreed to serve as chairman of the Imperial's board of directors, and several of his prominent friends joined the board with him. Vice-chair was Henry Syme of Gresham House, soon to become one of Chalmers's staunchest allies, as well as an occasional burr beneath his saddle.[32]

Although Chalmers's assay seemed to back up the mining company's prospects, the new slate of directors determined to "send out a competent party to examine the property, take possession, and if at all like the representations, push the tunnel forward."[33] That person, as it turned out, would be Lewis Chalmers. The Imperial's prospectus now dubbed him its "Managing Director at the Mines."

For Chalmers, it must have seemed like a golden opportunity: a chance to start over in a whole new country, with a solid £1,000 *per annum* salary to support his family while he was gone.[34] To be sure, he had no previous experience at mining, other than his recent study of assaying, but he was used to managing businesses and accounts. Even more, he was used to managing people. No doubt Chalmers was confident he could learn whatever else he might need to know about mining once his feet hit the ground at the mine.

Sketch showing multiple silver lodes and tunnel, from 1863 prospectus of the Michigan Tunnel Co. (later acquired by the Imperial Silver Quarries Co., Ltd.). (*Author's collection*)

Chalmers shook British soil from his shoes on September 11, 1867, leaving Liverpool, England, aboard the steamship *City of Paris*.[35] Chalmers's immediate aim, he frankly admitted, was to make money. He swore he would never return to England "unless I can show I accomplished my purpose."[36] On the one hand, that meant he was bound and determined to make the Imperial's mining venture a success. But the full truth was larger than that—his unstated goal was to recover his reputation.

The *Paris*, a new addition to the Inman steamship line, had been in service for less than two years when Chalmers stepped aboard. One of the fastest steamships at the time, she was later clocked at 13.75 knots per hour. Some 346 feet long with a clipper-style bow and a single-screw design, she featured not only a large smokestack but also three masts for sails. While an unknown number of passengers rode in steerage, Chalmers was one of the 143 who shelled out the cabin fee for the far-more-comfortable stateroom accommodations.[37]

The *Paris*'s journey across the Atlantic took eleven days, docking in New York City on September 22. Although a Pacific Mail Service ship departed New York for Aspinwall, Panama, on October 1, Chalmers was not on it. Instead, he boarded the steamship *Henry Chauncey* another ten days later, on October 11. His company would chastise him for the delay, but Chalmers explained he had been waiting for the tardy James Jones, who did not show up in New York as promised.[38]

The *Henry Chauncey* was a sidewheel paddle steamer, first launched in 1864 as part of the Pacific Mail Steamship Company fleet. None other than Samuel Clemens would describe the *Chauncey* as a "magnificent ship" just a few months later, when he, too, sailed to Panama on this same route in March 1868.[39]

After docking at Panama, Chalmers would have crossed the Isthmus by train, a quick four-hour journey, before boarding another ship for San Francisco. This was the *Golden Age*, the fastest of the Pacific Mail Steamship Company fleet, and on this particular voyage, she carried an astonishing 801 passengers.[40] Chalmers finally was able to step off the steamship at San Francisco on November 3, 1867, but his journey to Alpine was far from over.

Another steamship would have ferried him up the Sacramento River to reach Sacramento, the thriving capital of California. There he checked into the Orleans Hotel on November 7 for a good night's sleep.[41] Perhaps all too aware he was headed for a rough-and-tumble mining camp, Chalmers also stopped in at the storefront of Schraber & Davis to purchase a revolver and ammunition.[42]

From Sacramento, Chalmers may have boarded the Sacramento Valley Railroad for its newly-completed leg to Shingle Springs (just west of Placerville, California), a distance of about 35 miles. Once there, all comfortable traveling options would have ended and he would have been forced to catch a stage for the long and exhausting ride over the mountains.

Finally, on November 9, 1867, Lewis Chalmers stepped off the stagecoach on the dusty streets of his new home, the mining town of Monitor.[43] It was almost exactly two months after he had left Liverpool. Chalmers was forty-four years old—hardly a spring chicken—but he was off on a new adventure. Little did he know that his troubles had just begun.

The steamship Henry Chauncey, *circa* 1864. (*Courtesy of The Huntington Library,* image #pri JHK 00339)

The bustling port of Aspinwall (today's Colon, Panama), showing the train that carried passengers across the Isthmus of Panama. (*Harper's Weekly, 1868 May 30 (Library of Congress*))

2

Flying High with the Imperial (1867–1868)

His first glimpse of Monitor, California, must have come as quite a shock to Lewis Chalmers. In place of the rolling hills and heather-clad mountains of his native Scotland, he now found himself deep in the shadows of a narrow canyon, surrounded by jagged, rocky outcrops towering overhead. Instead of the vast, moody North Sea, here the only visible water was a small creek tumbling down the gulley of the canyon. Each breath he took would have reminded Chalmers he was not at sea-level anymore, but perched some 6,100 feet in the air.

Like Silver Mountain City, its sister mining camp, Monitor had blazed into existence during the silver excitement of the early 1860s. Inspired by the 1858 Comstock strike roughly sixty miles to the north, eager miners had fanned out throughout the mountains seeking similar riches. And, from surface indications in the croppings, at least, they thought they had found them here at Monitor.

In July, 1862, three miners working the cliffs south of Monitor Creek had set off a blast that uncovered ore assaying as high as $3,000 per ton. Predictably, a mining rush ensued. By the following May, some 249 claims had been staked near the new mining camp.[1] Like many such excitements, however, speculators were as quick to arrive as those working picks and shovels—and often those two groups were one and the same.

> During this unnatural state of things, hundreds of claims were located, duly incorporated, the stock issued, and thrown upon the market. The greenhorns bit like a hungry shark—everything was silver. They bled freely in assessments, not one dollar in ten of which was ever expended upon the claims.[2]

In March 1863, a formal set of mining laws was adopted for the Monitor Mining District, and by the following July, a township had been formed, complete with a constable and justice of the peace.[3] The name Monitor was bestowed on both the town and the mining district to honor the Union ironclad ship that had just won a crucial Civil War naval battle with the *Merrimac* in March 1862, "an event that so electrified the country, and which saved the nation from a great disaster."[4]

Nobody bothered to use similar creativity in naming the town's three major streets. There was, of course, a Main Street; this hugged the creek and was where

View of the town of Monitor *circa* 1871, showing the Tarshish Hotel (named after the nearby mine). (*Courtesy of the California History Room, California State Library, Sacramento, CA* (*Winchester Collection*))

Chalmers's stage likely deposited him. Paralleling it one block to the north was North Street, pressed against the foot of the mountain, while the equally-predictable South Street meandered along the south side of the creek.[5]

By the summer of 1864, the townsite of Monitor stretched nearly a mile down the canyon and featured some 150 buildings, with "timber for as many more on the spot, awaiting the sawing of lumber to finish them."[6]

The relatively small size of the town probably came as a surprise to Chalmers. Back in London, Capt. James Jones had likely painted the settlement as far larger than it actually was. In a prospectus for yet another of his Monitor mines, Jones would boldly assert that "[w]ithin a radius of five miles from Monitor Town there is a population of 10,000 engaged in the Silver Mines"—a figure inflated by at least a factor of ten.[7]

In truth, Monitor's population fluctuated wildly, depending upon the current success of its mines, or lack thereof. After the initial rush of '63 had died down, for example, the summer of 1864 saw only about seventy-five souls left in town. Just a year later, Monitor was described as "almost entirely deserted," with "not above twenty people in the town and probably not over fifty in the district."[8] When Chalmers stepped off his stagecoach in late 1867, however, the town's population had swollen again to 100 to 150 residents, thanks to promising recent discoveries just south of town.[9]

Just a few years earlier, the *Sacramento Star* had ridiculed Monitor's "log huts, clap-board shanties, tents and shake houses." But by the time Chalmers arrived, it was no longer entirely devoid of the trappings of civilization. A telegraph line, hastily strung across the mountains, had reached Monitor in June 1864.[10] Sherman G. Lewis, publisher of the local *Monitor Gazette*, had happily made room for the telegraph battery in his Main Street office—now serving triple duty as headquarters for the newspaper, post office, and telegraph. Editor S. G. Lewis himself, of course, also doubled as town postmaster.[11]

Regular stagecoach service to Monitor had commenced about three years before Chalmers arrived, with one stage line touting its eighteen-hour service between Sacramento and Markleeville, and connections from there to Monitor and Silver Mountain.[12] Both a hotel and stable had opened at Monitor by early 1866.[13] Markleeville butcher C. H. Kilgore also ran a regular meat wagon, keeping Monitor's residents supplied with fresh meat.[14]

Yet like so many early mining camps, Monitor remained fairly primitive when Chalmers arrived. The most substantial buildings were built of milled lumber, but log cabins still dotted the town. The fundamentals of drainage had been lost on some eager settlers, who built too close to the creek flowing through town. Just a month after Chalmers's arrival, a cloudburst swelled the creek, damaging some of these poorly-situated buildings. Any lessons learned from the experience seem to have been soon forgotten, however, as seasonal flooding would be a problem in future years as well.[15]

The earliest mining claims had not only dotted Monitor Canyon but also fanned out north and south along the Carson River at the foot of the canyon. By the time Chalmers came to town, the largest still-active mine was the Tarshish, a quarter-mile below town. Dubbed the "representative mine of Monitor" by Chalmers, the Tarshish had first been located about 1863, and in 1866 was snapped up by the Schenectady Gold & Silver Mining Co. of New York.[16] A fresh silver strike in their tunnel in spring 1867, six months before Chalmers's arrival, was still generating considerable local excitement.

Alpine County judge Henry Eno had ventured into the Tarshish tunnel the previous April to see things for himself:

> [The silver lode] was fully eight feet wide and the richest silver ore I ever saw. Two or three assays have been made as high as $500 and $700 to the ton. They are not through the ledge [yet] and don't know how wide it is. The telegraph operator here [at Silver Mountain] says he received a telegram this morning that the ledge was twelve feet wide and that ore that had been taken to Virginia City assayed $1,400 per ton. This strike in the Tarshish lode will create a great excitement and ... we may expect all the Mining Speculators, all the Mining Sharks and reckless adventurers to come prowling among us.[17]

The Tarshish was "the biggest thing in the mining line on this side of the mountains," similarly gushed the newspaper at Silver Mountain City, dismissing the mines of Nevada as "no where."[18]

Local excitement had been further ginned up that spring by production of a silver brick, smelted at the Ophir Mill in Washoe from 9,195 pounds of Tarshish ore. This tangible proof of the Tarshish's riches measured 10 × 4 × 4 inches in size and was valued at an astonishing $1,203.82. "Eight such bricks as this would reimburse the company the entire outlay" for its mine, boasted the *Alpine Miner*.[19] The entire town of Monitor celebrated the mine's success, convening a "grand ball" at which the massive silver brick was featured proudly on the music stand.[20]

A three-story ore house was quickly thrown up to hold fresh supplies of Tarshish ore, and by June 1867, some twenty-one workmen were busily engaged in hauling ore from the mine.[21] The Monitor Post Office (still housed in Lewis's newspaper office) proudly displayed samples of three different classes of the ore: one yielding a mere $20 per ton; another estimated at $125 per ton; and a third producing $400 per ton or, with 'washing,' an astonishing $3,000 per ton.[22] "Beautiful cabinet specimens" of rose-red Tarshish ore with its "peculiar crystalline form" were also passed around for curious viewers to admire—little wonder, then, that the town was in high spirits when Chalmers put in his appearance.[23]

A Rude Awakening

After months of hopeful preparation followed by weeks of tiresome travel, the gently-bred Chalmers was in for a rude awakening when he finally struck town that chilly November: no hotel accommodations were available, thanks perhaps to the Tarshish's recent success.

Putting his celebrated powers of persuasion to use, Chalmers somehow convinced newspaperman Sherman Lewis to let him board with the Lewis family—hardly a comfortable arrangement, but "the only place I could get," Chalmers grumpily wrote home.[24]

Nothing went quite as expected, either, when he sought directions to his company's mine. The project he had come all this way to manage was not actually in Monitor proper; it actually lay almost 2 miles down the canyon, near an even smaller encampment beside the East Fork of the Carson River known as Mount Bullion.

Worst of all, after Chalmers trudged determinedly down the canyon and presented himself at the tunnel to officially take possession, he was greeted by Michigan Tunnel Company secretary Oscar Thornton, who flatly refused him entry.[25]

Chalmers may be forgiven if some shouting and cursing followed, though he did not record that reaction in his letters. What he did record was his unhappy discovery that the London syndicate he represented actually did not own the real property on which the tunnel sat. The company's purchase agreement with the Jones brothers in February 1866 had been for stock of the Michigan Tunnel Company.[26] Yet unbeknownst to the London investors, six months after that deal had been inked, a judgment against the Michigan Tunnel Company by unpaid workmen had resulted in a foreclosure sale. In October 1866, nearly a year before Chalmers left England, the Michigan mining property had been auctioned off at a sheriff's sale to satisfy that judgment.[27]

The Jones brothers had not breathed a word of this little hitch to the buyers of their stock back in London, of course. So it was only after Chalmers arrived in Alpine County that he discovered that title to the property was now held by a former contractor named Petiss, who had purchased it at the sheriff's sale.[28]

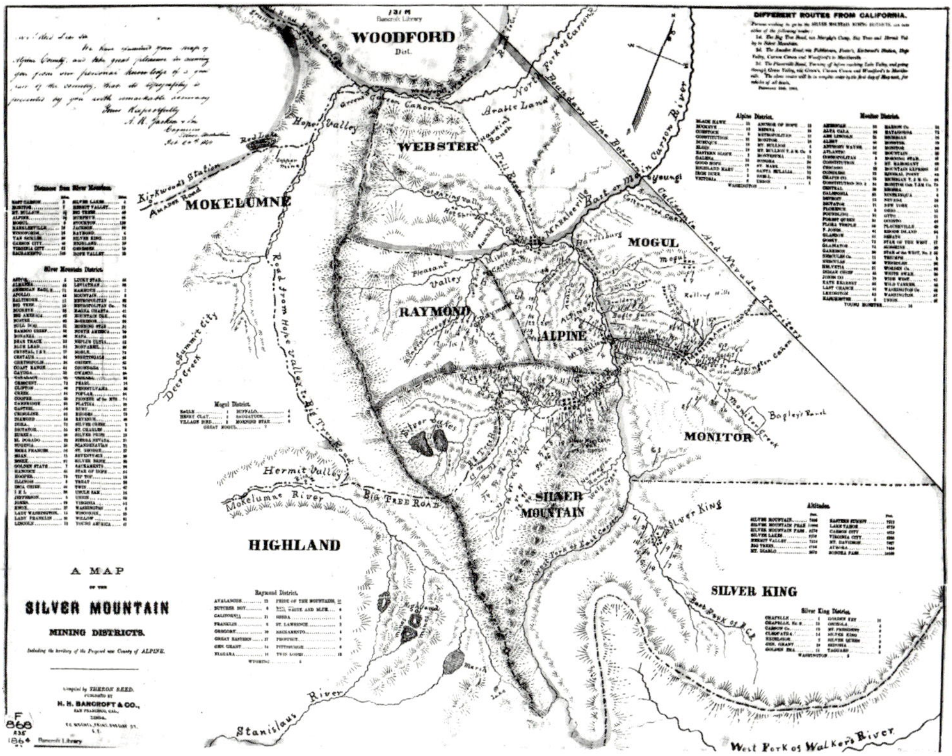

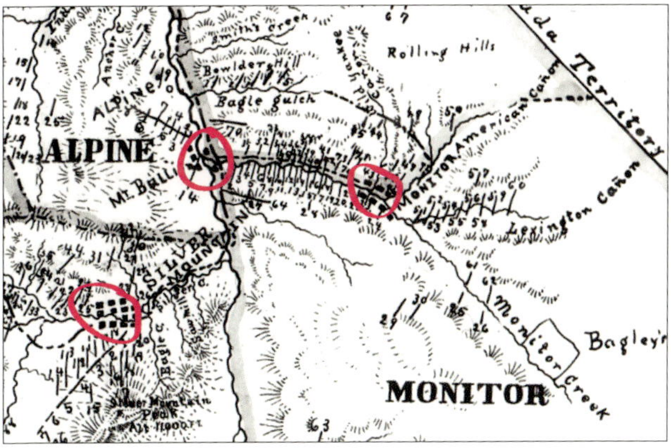

Map of the Alpine County mining districts in 1864, with mine locations indicated by lines with small numbers. Inset shows the location of the towns of Monitor, Mount Bullion (Bulliona), and Silver Mountain City. (*Alpine County Historical Society*)

In yet another wrinkle, Chalmers also learned that Joneses were not the owners of all of the shares of the Michigan Tunnel Company for which they were demanding payment. Some of the company's shares had been casually "allotted" to others on the corporation's books, without stock certificates being formally issued. So even completing the acquisition of the Michigan company's stock meant locating assorted third parties, who would then need to be satisfied.[29]

Thoroughly hot under the collar by now, Chalmers determined to get to the bottom of things. He quickly convened a series of meetings with Thornton, Petiss, and Frank Jones and, lawyer that he was, threatened to sue them for fraud.[30] Chalmers also dispatched a hasty cable to London, advising that the Joneses were demanding the full price promised for their now-worthless stock and warning his directors to withhold shares of the Imperial Company that had been promised to the deceitful John Jones as part of the deal.[31]

The problem, of course, was that Petiss was a lawful buyer at the foreclosure sale, and the Joneses and Thornton were so poor they were judgment-proof. Even if the company won such a lawsuit, Chalmers warned his London directors, "we could not recover."[32]

The Joneses had an entirely different view of things, of course, and fingers were soon pointing on all sides. James Jones claimed he had sent $5,000 to his brother, Frank, to pay off Petiss's judgment.[33] Frank Jones contended he had never received the money. And both of them thoroughly detested Chalmers. "I have only had four confabs with [Chalmers] and I do not like him," Frank snarled to his brother James in a letter that December. "I think he is a shark."[34]

The dark feelings were mutual; Chalmers did not care for the Joneses or their associate, Thornton, either. "Frank Jones is cunning and tricky, O.F. Thornton an out-and-out scamp, and I wouldn't now trust Capt. Jones unnecessarily," he advised London, adding that Sherman Lewis, his new landlord, shared his poor opinion. "The Lewises and Joneses have been at feud for years," reported Chalmers.[35] He soon came to suspect that even Petiss, the nominally innocent buyer, had actually been a "tool" for Frank Jones and Thornton.[36]

Yet when it came to obtaining possession of the tunnel, Chalmers's hands were tied. There was nothing for it but to pay off Petiss to clear title to the mine.[37] It took over two months, but things were eventually smoothed over. Chalmers applied pressure to Frank Jones, who was forced to admit that his brother, James, had indeed sent him $5,000 to redeem Pettis's claim. Frank finally handed that money over to Petiss to satisfy his claim, and the property was transferred by Petiss directly to the London buyers. Finally, on February 16, 1868, Chalmers was able to take possession of the mine he had come so far to manage.[38]

An initial report submitted to the Imperial's directors on February 4, just before Chalmers took possession, had contained the Jones brothers' estimation of the Michigan/Imperial mine. The tunnel, it predicted, would cut "no fewer than forty silver mines," with "enormous and each month increasing profits," enthusing:

> Mining generally is no longer an adventurous [or] a speculative pursuit, if experience and honesty of purpose are combined ... The experimental failures of 50 years ago have, necessarily, paved the way for the brilliant successes of the last half decade.[39]

Investors an ocean away had been led to believe that the Michigan tunnel was already within 50 feet of its first ledge (the Fashion); would soon cross a second lode (the Senate); and was within 150 feet of yet a third (the Triumph). They had also been assured that the Fashion measured 8 feet wide and would produce 73,000 tons of ore, and that shipments of ore could commence "about the middle of May [1868]."[40] What James Jones had not shared with the London buyers was that as recently as September 1864, his brother Frank had written him: "As to the Michigan Tunnel, I think that it is a complete failure."[41]

But when Chalmers took over, the Michigan Tunnel—now grandly renamed The Imperial Silver Quarries Ltd.—bore little resemblance to the heady tales spun in London. In place of the 279 feet of tunnel that James Jones claimed had already been run, the tunnel actually stretched just 180 feet into the hill known as Mount America, and even that work had not been done to Chalmers's satisfaction.[42]

Chalmers wisely distanced himself from Capt. Jones's previous predictions, advising the directors, "I think it my duty to say that I am not to be considered as in any way committed to [Jones's] Statements, more especially as to first shipment of ore, time of reaching ledges, and distances from mouth of the Tunnel," adding politely that Jones, having left at the height of the mining excitement, "is apt to color too warmly."[43]

Nevertheless, Chalmers set to work with a will. Determined to waste no more time, he signed a contract with miner William Mercer of Silver Mountain just one day after taking possession. Mercer was to drive an additional 150 feet of tunnel at $12.45 per foot, not including track, timber and ventilation.[44] Deeming the

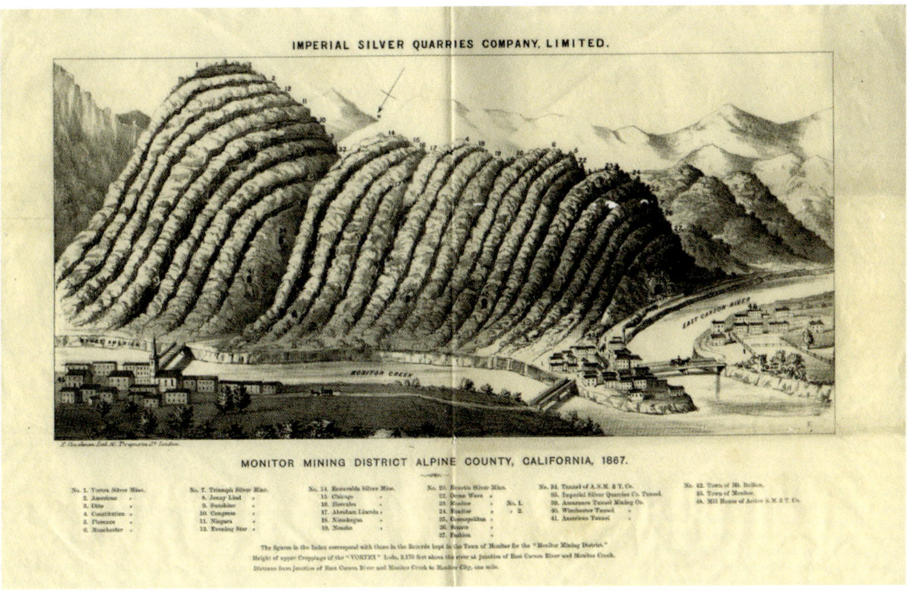

A charming and fanciful sketch from the Imperial Silver Quarries prospectus, showing the Imperial lodes and other mines in the vicinity. Capt. Jones was not the only one to exaggerate; the buildings and improvements shown were purely imaginary. (*Michigan State University Archives and Historical Collections, James Jones papers, collection c.103*)

existing headroom in the tunnel insufficient, Chalmers also set a crew to work "heightening" the old tunnel to bring it to a full 6 × 7 feet in size.[45]

The tunnel itself was surprisingly large for an untested mine, wide enough for a double set of rails so incoming and outgoing ore cars could operate simultaneously. The floor sloped 7/8 inch in every 16 feet for drainage, and a 12-inch-deep trough was carved in the bottom to carry away water.[46]

Under Chalmers's watchful eyes, work in the tunnel was pushed "night and day," with contractor Mercer's men at work every day of the week except Sundays.[47] Even so, progress was not without hiccups.

The rock grew harder the farther in they got, and Mercer began having cash-flow troubles. At least one of his men sued him. Then, as the end of the agreed-upon 150-foot goal finally approached, "the running got so bad that the men left ... and [Mercer] had to abandon his Contract," Chalmers reported to his backers in London. Luckily, the canny businessman had held back a portion of Mercer's pay against just such an eventuality. Although Mercer had been promised $12.45 per foot, "I finished the Contract [from the retention]," Chalmers confirmed, "at a cost to him of $20 per foot."[48]

On March 25, an unexpected bolt from the blue rocked Chalmers to his core. A pair of letters from sister-in-law Eliza, who was caring for his children, reported that his eldest son, Lewie, now fifteen, had tuberculosis. In retrospect, the shaken Chalmers confessed that he had "noticed a whoop sometimes," but taking his family's good health for granted, he'd given it no particular thought. He dashed off a solicitous letter to Lewie:

> I cannot tell you how much shocked I am ... Ere this reaches you, however, I do hope and pray you will have so far recovered as to be able to get out, although you must not despond if such an illness as you have had leaves you weak for some time. Lessons you must not on any account attempt, obey implicitly the Doctor's prescriptions, and if the funds are forthcoming, [Aunt] Eliza will have arranged that you have a change of air, which will be of more use to you than anything else.... Get yourself a fishing rod and go where you can get some hunting, always consulting the Doctor as to the climate &tc.... One thing I must impress on you, and I cannot do so too strongly, and that is, don't get down hearted—very few indeed there are whose lungs are unaffected; you have caught cold, somehow, and your system was weakened after the fever, you may have studied too closely, and growing as you have been, all has been too much for you. But care and attention (and I am sure you will have both from Eliza) will, with God's blessing, I hope, restore you again to perfect health.[49]

Separately, Chalmers wrote to Eliza, urging her to get all the children, and especially Lewie, out of London:

> You may fancy the uneasy feeling I have about Lewis ... Scotland is the place for him to go, and as in all cases like his, mind as well as body ought to be consulted.... Let him have a fishing rod (not too expensive) and fishing to amuse him. Meantime give him goat's milk once a day or oftener if he can take it. (I enclose an Article on the healthfulness of Apples. Let the children have them in the morning occasionally, if they like them for Breakfast, with Bread and Butter, no Tea.) A little cod Liver Oil

> when any of them droop would be good, but above all, plenty of air and exercise.... If you have means, take the children to Scotland, by steamer or rail as you please. Wouldn't you like Dunkeld better. Lewie would have fine fishing there on the Lochs, and Ellie was always better [for] her trips there. Besides, you would be unknown and wouldn't hear tattle.[50]

Fresh missives soon brought the happier news that Lewie seemed to improving, and Chalmers changed his plans. Now, he proposed to send Lewie to a Dresden school for "Mining, Military & Government appointments." There, he could learn "the best German," associate with companions from influential families "which may prove useful in your after life," and enjoy a "delightful climate" beneficial to his health, he urged Lewie. "I have carefully thought over all this, and am satisfied that no profession now offers so much chance of acquiring a good position as that of Mining Engineer." At the same time, however, he gently inquired about "your own views and inclinations."[51]

By early May, Chalmers was privately confessing that the tunnel's first two lodes were disappointments. "[T]he so-called Senate and Fashion lodes are not well-defined; indeed, I begin to think they are only surface stains, or float rock from the Triumph," he wrote his brother. "I do not ... anticipate finding ore worth Shipment, if any, until I reach the Triumph." And that, he estimated, would require at least another 400 feet of tunnel.[52]

Nevertheless, he urged the Imperial's board not to lose faith:

> [T]hey must not be impatient or discouraged because the results are not so speedy as Jones led them to expect. The first part of the work will be the most tedious—the lay of the mountain is deceiving, but time and money will accomplish all I have said; and that is worth waiting for. Once at the Triumph (if indications are worth anything) we shall not only have a large body of ore but we shall be so far under the surface that we may look for fresh discoveries at short intervals; all of which I will duly telegraph.[53]

By late May 1868, the first 150 feet of new tunnel had been successfully completed, a major accomplishment given the hardness of the rock.[54] Chalmers was even winning some respect from Oscar Thornton, his former adversary. In a letter to a mining friend, Thornton privately acknowledged that Chalmers had "finally proved Capt. Jones and his brother to be very bad men."[55]

Finding it impossible to hire an additional stretch of tunnel run for a reasonable sum, Chalmers now determined to push the work himself. And push it he did in a most ambitious fashion. He hired seven men, setting three two-man shifts to work around the clock, with a seventh man dedicated to sharpening tools. Using this arrangement, he soon was making nine to eleven feet of progress a week in the tunnel.[56]

That was not quite all Chalmers had been doing, though he did not broadcast his side activities to the directors. He would later admit to his brother that in January and February, while he was waiting to take possession, he had managed to pick up two mining locations for himself, paying an old miner to do the required five days of labor on each to hold the claims.[57]

Shabby is Not in my Composition: Building Imperial House

Though the Imperial Tunnel had not yet produced its first ounce of pay ore, Chalmers set about expanding its assets. In May 1868, he purchased 160 acres, a mill site, and a water privilege beside the Carson River for the company.[58] A site for a mill certainly seemed a premature investment for a non-producing mine, but when that first silver lode was reached, Chalmers wanted to be ready. There was also a personal benefit to this seeming madness.

The nearly 2-mile trudge between Monitor and the mine and back again proved most unwelcome exercise, and boarding with the Lewis family had become tedious. "[T]hey have raised my Board lately and washing is very dear," Chalmers complained.[59] But lo and behold, the mill site he had purchased for the company was directly across the river from the tunnel's opening, a spectacularly convenient spot for a manager's house.[60]

A house close to the tunnel was a necessity for proper supervision, he advised the company in May 1868, and building one was the only viable alternative. "I find there is not [another] place of any description to be got."[61]

The spot where he proposed to build sloped steeply down to the river and required considerable excavation to create a buildable pad, but Chalmers had crews at his disposal. In June 1868, he diverted two shifts of miners from the tunnel to hack out a flat spot in the hillside for a new house, office, and assay office. The grading work alone consumed six days, an expense he charged in his accounts to "Mill Site."[62]

"I leave the Lewises so soon as my wooden Shanty ... is completed," he advised Syme. The cost, he blithely assured the chairman, Lord Poulett, "[w]on't exceed $600 ... Besides, it keeps the ground for the Mill Site."[63]

His "shanty," as Chalmers called it, was hardly a diminutive one; the footprint was a comfortable 37 × 21 feet and it rose two stories tall.[64] To the Imperial directors, he described it simply as a "house containing four rooms, an office 15' × 20', parlor, bedroom and kitchen," to which he had "added a wing containing Assay office and laboratory, at an additional cost of about £95 Sterling."[65] But this quick summary understated things a bit. In reality, the house featured a kitchen on the ground floor with an office and parlor above, each 15 × 20 feet in size. Outer porches wrapped around the tall structure, and not one but three woodstoves helped ward off winter chills.[66]

Chalmers had been living frugally, he assured his sister-in-law, after she accused him of being "shabby" (evidently for failing to send home as much money as she expected). He wore each shirt and undershirt for a fortnight to save money on laundering, he protested, "never tasting Spirits, beer or wine," and giving up smoking entirely. "[Shabby] is not in my composition," he growled.[67]

When it came to his new house, built with company money, Chalmers proved not shabby indeed. Running water was supplied to the house by a pump. Windows stretching nearly 6 feet tall—more than a dozen of them—afforded fine views of the river and the small town of Bulliona in the distance.[68] Outside, the structure featured two separate doorbells, one for the house and one for the office, plus foot scrapers to ensure dirty boots would not track in mud and debris. Hardware for the doors included porcelain handles and silver-plated key plates.[69]

The corner of Chalmers's "Imperial House" is shown at left in this rare photo, with Coulter's dam below, and the town of Bulliona in the background. Photo is believed to have been taken by P. W. Greenfield on 6 August 1871. (*Courtesy of the California History Room, California State Library, Sacramento, CA* (*Winchester Collection*))

By late July, Chalmers's house was "nearly inclosed," and in early August, shingles to finish off the roof were requisitioned from Cary's Mill at Woodfords.[70] Chalmers was finally able to leave Monitor and move into his new home around mid-August, and expected to finish the adjoining Assay Office "in a fortnight or so."[71]

No longer willing to squeak by without domestic help, he procured a "married miner and his wife" to keep house for him in exchange for their board.[72] He also indulged in a touch of vanity, ordering two bottles of "Hall's Hair Renewer"—a coloring product. "Every man here is bald—be he young or old. There is something in the climate prejudicial to luxuriant hair," he wrote to his family, admitting he was going grey and now had no hair on the top of his head.[73]

Chalmers placed a fair-sized order with a merchant in Sacramento for furnishings for his new abode. Yet again, he seemed determined to make himself as comfortable as possible in this strange, wild place. On his shopping list were not one but two wine decanters, cane-seated chairs, and ivory-handled knives. For local miners, these might have counted as luxuries. But for Chalmers, they must have felt like necessities.[74]

Finally finding himself reasonably comfortable, Chalmers encouraged his younger brother, John, to come out:

> The climate is good, but you have to wait on yourself. Grub fair, but dear, brandy and whisky bad, bac [tobacco] good. I have dropped both grog and pipe and feel much better without either.... Woman servant $50 p[er] month. Grass widows, fond of men.[75]

Examples of small assaying cupels, on display at the Alpine County Museum. Chalmers kept a small, adjoining assay office at his house and did his own assaying. (*Dustman photo*)

The cost of his new house came in higher than anticipated. Chalmers blew through his initial $600 estimate and in January 1869 was notifying London that another $600 would be required for "extra work on [the] house—to make it habitable." Upon protest from London, he eventually agreed to shoulder most of the additional expense himself.[76]

None of his recent improvements were met with warm approval in London. Although the company grudgingly sent an additional $5,000 in July 1868 to keep operations going in the mine, Chalmers also received a telegram sternly admonishing him to "limit expense to tunnel."[77]

With a house to call his own, Chalmers finally enjoyed a bit more privacy, but quickly discovered that life in Bulliona came with its own annoyances. He shot off a pointed letter to one neighbor, expressing hope not to be thought "unneighbourly or disobliging" but requesting that the man keep his horses off the property below Chalmers's house "because it destroys the run for my chickens and ... there is ... really no food for horses."[78]

His commute back and forth to the mine now involved only a matter of minutes. But with no bridge across the river, Chalmers was forced to step his way across on a "rickety foot log."[79] Worse yet, the insubstantial arrangement was "apt to be carried away when the River rises." Chalmers began pushing his London backers for funds to build a "viaduct"—a bridge that would allow the miners to dump waste rock on the far side of the river. The ostensible goal was to enlarge the future mill site. Yet a bridge would, not coincidentally, make it much easier for Chalmers to get back and forth. A flume, too, was part of his new proposal.

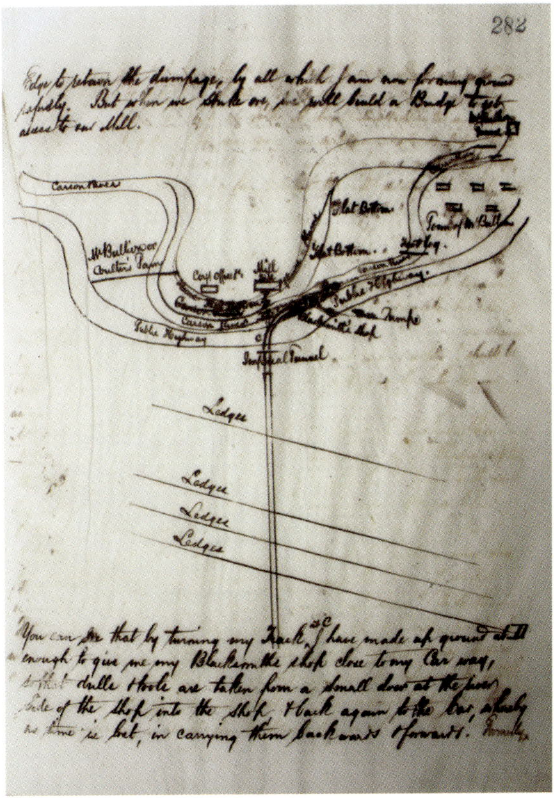

Chalmers's sketch of the Imperial tunnel cutting its anticipated ledges, with his plans for proposed improvements across the river, including a house ("Company office"), mill, and flume. The town of Bulliona is shown at far right. (*Chalmers Letters collection, Alpine County Historical Society and Museum, 1869–1869 Personals, p. 282*)

Such high-flying ideas were further evidence of Chalmers's signature extravagance, according to Bulliona resident Frank Jones. Writing to his brother, James, Frank sneered: "Chalmers intends building a costly bridge … So goes the money."[80] Brother James ran with the gossip, penning a pair of tattle-tale letters to two Imperial investors in December 1868, complaining about Chalmers's use of tunnel workmen to grade for his own house and his latest "rather expensive" plan for a bridge over the river.[81]

Pushback from London intensified, and the Imperial Board took Chalmers to task over his expenditures. But Chalmers remained unbowed. "[I]t was nearly impossible to live at Monitor and do my work at Mount Bullion," he retorted, and "it was necessary to hold the ground for [the] Mill Site by personal occupation." In any event, insisted, he had "saved you $400" by building a new house because he "could find no dwelling place" at Mount Bullion. The only available alternative, he contended, would have cost the company even more: buying a nearby house at its $1,000 asking price.[82]

Progress and Personalities

Although he brought no previous mining experience to California with him, Chalmers did have years of handling business affairs, and that background helped him avoid at least one costly mistake. Long before Chalmers had left London, the Imperial's board of directors had acquired an Irish fifty-stamp water-powered mill, rail, amalgamating machinery, and wagons, and had shipped them all the way from Liverpool to San Francisco. There they had been rusting at the dock, accruing $6,000 in unpaid duties and storage fees.

Concerned about the high cost of freighting 140 tons of equipment over the mountain sight unseen, Chalmers wisely arranged for mining expert Guido Kustel to examine the equipment to determine its condition. That bit of caution proved well-placed. Kustel reported that "considerable" damage had been done to the machinery in transit. To the board's chagrin, the well-intentioned mill was not worth the trouble and expense of getting it to Alpine.[83]

As the Imperial tunnel work progressed, Chalmers was forced to learn a great deal about mining, much of it the hard way. When the tunnel began requiring better ventilation, for example, he began searching for blowers to improve the air for his men. He wrote B. F. Sturtevant himself in Boston, Massachusetts, requesting information on the popular Sturtevant Blower.[84]

An early adaptor by nature, Chalmers also began researching the latest developments in mining technology. He expressed great interest in the "famous Burleigh Pneumatic Drill" attracting attention in *Mining and Scientific Press*.[85] He also ordered a keg of Giant Powder, a newly-developed form of dynamite being touted as a significant improvement over black powder explosives. "If it accomplish half of what it claims, it will be a great boon," he wrote to his brother, John. Although Chalmers's first experiments with the new blasting powder were unsuccessful, later ones proved its worth. He did complain about its cost to the supplier, though. "[B]efore you get it generally used, you will require to lower its price," he snipped. Before long, however, he was touting the product to London as "invaluable" in wet mining conditions.[86]

Chalmers also decided to try his hand at photography, and in July 1868 had a complete set of photographic equipment shipped "in two heavy cases" from Messrs. Ross, Opticians in Cavendish Square, London. Photographic supplies were scarce, however, and weather conditions were against him. Among other hurdles, his developing solutions kept freezing solid. Finally, he cussed, "Photographs. Hang them. Every step I move, I find you have omitted to send me something." Still, he did not give up. He ordered several books to help him learn the technique and, with practice, was eventually able to send photos back to London.[87]

Determined to serve as his own surveyor, Chalmers requested that a transit be sent out.[88] Surveying was apparently a skill he already possessed, and he would continue to survey various claims and tunnel projects in the years to come.

Chalmers also put his training as a lawyer to use when the local tax assessor tried to value the Imperial's as-yet-unproductive mine at a whopping $1,500. Since Alpine County officials were paid in county scrip, Chalmers explained to the folks back home, they "have suddenly become alive to the fact that a good thundering assessment on the Mining claims would materially advance the value of that scrip."[89]

Chalmers "entered into a defensive alliance" with other mine owners to contest the exorbitant mine valuations. Observing that he "wouldn't give a fig for the opinion of the Lawyers here," Chalmers paid for an opinion from legal expert Gregory Yale, author of a book on California's mining laws, to support his view that the proposed valuation was far too high.[90, 91] He also held forth for six hours at a county supervisors' meeting, protesting the unfairness of high taxes on an unproven mine.

When even that did not seem to do the trick, however, Chalmers joined with the other mine owners to quietly pay a $50 bribe to cut their tax assessments in half. This, he wrote home, proved to be "the Golden Key."[92] The expensive expert's tax opinion had been of little use, "bribery more convincing to the minds of the 'Honorable' Supervisors of the County of Alpine."[93] Chalmers would later press his arguments in a lengthy letter to *Mining and Scientific Press*, renewing his complaint about the unfairness of taxing undeveloped mines.[94]

All in all, Chalmers assured Eliza, he was working very hard, "fitting up my assay office, attending to the Men and business of the Mine," while his evenings were quietly spent "consumed in reading."[95]

Ill feelings between Chalmers and the Jones brothers continued to fester, and to Chalmers's chagrin, his erst-while supporters in London seemed to put more stock in the Jones brothers' negative reports than his own.[96] Worse yet, Frank Jones now had his eye on Chalmers's position, and rumor had it Jones had been "promising men work in the Tunnel."[97]

A frustrated Chalmers cautioned Syme that James Jones "has the character of being the biggest liar [who] ever came to Monitor," adding in a later letter home: "You are made of different stuff from what I supposed if, on his first appearance at your tunnel, you did not make him more intimately acquainted with the point of your right boot and [the] Carson River."[98] Chalmers was on no better terms with Oscar Thornton, whom he dubbed "mine enemy." Thornton may have been "the smartest of the gang," Chalmers wrote, but "I am old enough now to play with edged weapons without cutting my fingers."[99]

Imperial Indisposition

With no silver to show for its efforts by May 1868, the Imperial Company began having difficulty attracting new investors, and its finances were stretching thin. The original agreement had been to cover Chalmers's traveling expenses and pay him £1,000 a year, "to commence from the day I left England," with a guarantee of at least £600 "if the company should turn out a failure." Now, however, the company tried to back-pedal, proposing to start his salary only from the date he had arrived in Alpine, without travel costs.[100] Chalmers protested loudly, and eventually won the salary battle.[101] But he recognized the dispute for the bad omen it was.

By August 1868, he was writing to daughter Laura that "Imperial indisposition upsets my plans," and privately acknowledging to Syme the possibility that the company might "go to the Wall."[102] Despite the financial troubles he was facing, Chalmers seemed to be enjoying his time in this strange and wonderful new land. He wrote warmly to his son, Lewie:

> I killed a little water snake in the Tunnel the other day, very pretty, striped in alternate bands longitudinally of black and gold, [and] one of my workmen killed near the office a largeish rattlesnake, three feet long and eight or nine Rattles. I often see beautiful lizards, rich green in colour, and the butterflies are beyond anything you can fancy. Grapes, apples, apricots, plums, peaches, [and] nectarines in great abundance, brought up from the Valleys in wagons almost every day. I wish you were all here. London cannot compare with this, notwithstanding the coldness of the winter.[103]

In September 1868, nearly a year after Chalmers arrived, the tunnel did finally cut a band of "Monitor coal," or obsidian—a welcome change from the "eternal porphyry," though equally worthless from a mineral standpoint.[104] But as for reaching the elusive Triumph lode, hard rock had meant slow running. Chalmers tempered his earlier expectations, predicting now that he would not reach the Triumph before February or March.[105]

Meanwhile, a continent away, the Imperial's directors were continuing to second-guess his management of their mine. Much to Chalmers's annoyance, some were accusing him of everything from "employing improper men" to "running the tunnel the wrong way."[106] With money a continuing problem, a petition to "wind up" Imperial began to circulate in England.[107]

In November, with no money to cover payroll, Chalmers wrote pointedly to Syme, demanding: "on receipt, it will be necessary that a supply of the needful be telegraphed." He also shot off a short letter to the Bank of California to inquire what their overdraft charges would be, "in case I should require to overdraw for a month or two, say $1,000 or $1,500."[108]

Nevertheless, to sister-in-law Eliza, Chalmers professed his confidence in the Imperial 'unshaken.' "Every day something transpires to augment my faith in this Mining Region," he wrote. "The Morning Star, and Tarshish, and I.X.L. are turning out rich ore—and so will I, by and bye."[109] He penned a similarly encouraging report to *Mining and Scientific Press*, confirming that the Imperial's tunnel was now 610 feet long, and describing the rock thus far uncovered.[110]

Christmas Day 1868 arrived, and although most of the crew were given the day off, Chalmers's foreman remained hard at work, "laying track &tc. &tc."[111] But as the year dragged to a close, Chalmers was decidedly gloomy. Despite a $2,403 remittance received in early December, he was short of money again for the upcoming payday, and described himself as "the poor devil they sent out to ruralize at Alpine."[112]

The precariousness of his own position was becoming all too clear. But what were his choices—return home and admit yet another defeat? That would have been especially embarrassing as he had passionately touted the Imperial's prospects to his friends. Giving up, like shabbiness, was not in his makeup.

That was not the whole of it. Chalmers had seen enough rich ore from nearby mines in Monitor and Silver Mountain to be genuinely bitten by the mining bug. His own assays of Buckeye ore for his occasional nemesis, Oscar Thornton, had yielded results as high as $357 per ton in gold and silver.[113] If his fortune was not to be made in the Imperial, perhaps other, more promising mines could be had.[114]

The canny Chalmers began casting his eye around for fresh possibilities. For once, he seemed willing to divorce his interests from those of his erstwhile friends in London, putting out a quiet feeler to an acquaintance in London. "I can pick up here one or two good undeveloped Mines.... Can you do anything with your friends?" he wrote one, adding: "Do not mention this to Syme."[115]

Then, too, with the Imperial still limping along, there was always the chance that his London investment group could diversify. Perhaps his bread was best buttered by continuing to work with the moneyed investors who had sent him; that, of course, simply meant finding another mine to "float" in London.

Chalmers had already approached the Imperial's board about purchasing the I.X.L. Mine. This was one of the earliest and most productive silver claims at Silver

Mountain's Scandinavian Canyon, and its owner, Daniel Davidson, had seemed eager to sell. As recently as August 1868, Chalmers had written to his friends in London:

> I can ship ore to you in a fortnight from commencing operations, or silver bricks if you prefer them.... I.X.L. ore is the finest I have seen in this Country, with the exception of the Tarshish, but specimens of the I.X.L. (selected) are finer than those from Tarshish, selected.[116]

His London contacts initially seemed open to the idea of acquiring the mine, but moved with glacial slowness. In December 1868, Chalmers wrote flatly: "You are too late." Two of the I.X.L.'s trustees had already purchased it and were, he said, hoping to freeze out the minority shareholders to keep the promising claim to themselves.[117]

Nevertheless, that same December, Chalmers was able to send home a proposal to buy a different Scandinavian Canyon lode: the Buckeye No. 2. Surprisingly, it was a deal he had hashed out with his sometime-enemy, Oscar Thornton. The clincher: it was a no-money-down deal, with the sellers receiving just stock in the new company and the promise of $12,000 from the first $17,000 produced in the future.

"I have examined [the Buckeye] very thoroughly, and had the benefit of the experience of Coulter, ... an old miner, who likes it better than anything he has seen here," Chalmers enthused to Syme. As he envisioned it, the Buckeye would be worked by a new company, to be called the Exchequer. "Try and get me the managership," he urged. "I can attend to it, and save the expense of sending another."[118]

If the Imperial were to crater, Exchequer was another iron in the fire. If push came to shove, this new mine and new company just might help keep him going in a whole new direction.

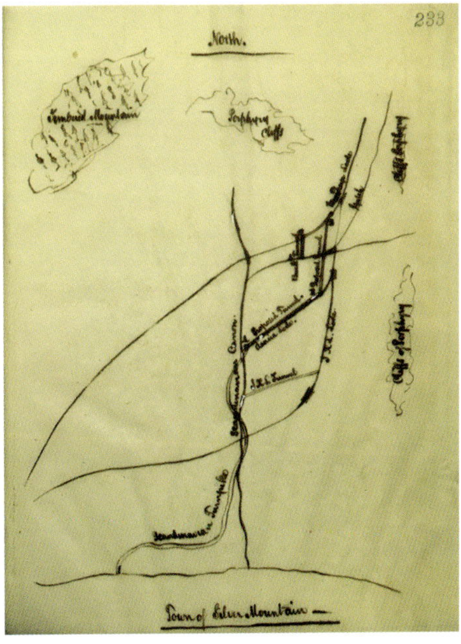

Chalmers's sketch showing the locations of the I.X.L., Buckeye No. 2, and Accacia Lodes in the Silver Mountain Mining District. At center is his proposed tunnel to tap the Accacia Lode. The I.X.L. tunnel and Buckeye Tunnel are shown as straight lines. Scandinavian Turnpike (a toll road) winds up Scandinavian Canyon, with porphyry cliffs on the top and far right, and a "timbered mountain" top left. (*Chalmers Letters collection, Alpine County Historical Society and Museum, 1868–1869 Personal letters*, p. 233)

3

The Imperial Craters (1869–1870)

Money troubles aside, as January 1869 dawned, Chalmers had a great deal of progress in the Imperial Gold & Silver Quarries tunnel to brag about, and brag he did.

Now a full year into his duties, he had seen the Imperial's original 180-foot tunnel grow to 490 feet in length, putting the working face some 460 feet underground. His crews had "widened, heightened and relaid track" in the old tunnel. A stove connected to a 58-foot-tall wooden chimney had been installed at the tunnel's entrance to help draw foul air away from the tunnel's face, and metal vent pipe had been tacked to the tunnel's roof with sturdy iron brackets, neatly "painted against rust." Water draining from the tunnel mouth had been routed to a tank in the blacksmith shop on the dump below for the smithy's convenience. Best of all, Chalmers's five-man crew was continuing to make between 9 and 13 feet of progress per week at just $11.91 per foot—far cheaper than a contract would have cost.[1]

The rock had not made for easy running. The first 552 feet of tunnel had been driven through hard porphyry, followed by a 2-foot vein of obsidian, and then a stretch of feldspathic lava "hard as adamant."[2] "There is no such obdurate rock in the Country as that through which I am now driving," Chalmers grumbled to the Imperial Board. "It drills badly, and blasts worse."[3]

His assays had picked up minute traces of gold and silver, "but nothing startling enough to send home." Sadly, that tiny bit of encouragement came from surface croppings rather than ore inside the tunnel.[4]

By now, money was tight enough that Chalmers confessed he had "given up all ideas" of the expensive mill, bridge, and viaduct. He even offered to slash his own salary until the tunnel struck good ore, if that would permit the company to keep going.[5] Once again, there was no money to cover January's pay day. Chalmers advised Syme that he was on the verge of having to stop work, and if the work stopped, he reminded him, "all you have sent is not worth a dime."

> I ought to have paid them off on the first, but trusting to you for a remittance in time, I have kept them on. If, however, on receipt of this, no money is forthcoming, telegraph "Stop work."[6]

Artist's illustration of a man pushing a wooden ore car. (*Scribner's Monthly, Vol. XV No. 4, p.481* (*February 1878*), *author's collection*)

Back in London, board members were again accusing Chalmers of mismanagement, but Chalmers merely pointed to the lack of capital they had provided. If adequate funds had been forthcoming, he argued, he would have been able to do more tunneling and sinking, exploring multiple ledges at once.

> Ten chances to one, had my suggestions to this effect [in my Report] been given effect to, I could today have been shipping ore.... With sufficient Capital, the Capital named by me in that Report, this is a first-class property. With inadequate means, like the rough diamond in the hands of a man who can't raise money to cut it, it's not worth a cent—and I am not responsible for the consequences. Your letters, after receipt of my Report, were so encouraging, 'no fear of Capital &c.,' that I put everything in excellent order, secured water power and Mill Site, built house and office, arranged everything for a permanent and prosperous undertaking which, with that Capital, it would be. Whereas, had I known how things were to turn out at home, I might have adopted a different and more hand-to-mouth sort of a course. Still, as I said before, I believe £500 will see me at the Triumph.[7]

Privately, Chalmers acknowledged serious concerns to his brother. "*Entre nous*, from the way Syme and you write as to money matters, the Company's days are numbered unless the Triumph chooses to come nearer than it has been for some thousand years."[8]

James Jones continued to add to Chalmers's troubles, this time by circulating a "calumnious circular" accusing Chalmers of poor management, self-dealing and waste. Meanwhile, James's brother, Frank, was confidently boasting to friends in Monitor that Chalmers would soon be discharged from his post. Frank did his best to aid that along by collecting affidavits from former employees, swearing Chalmers had been "the hardest, meanest man ever came in to the camp."[9]

An outraged Chalmers considered suing Jones for slander.[10] He protested to his London backers that such charges were simply a "foul calumny":

> I have the name here, and perhaps not undeservedly, of being the hardest taskmaster in the County. I am called mean (because I watch the men at work, not *the* thing here, you

> must know, where all are, or would be, gentlemen Miners, Superintendents, and men on a par); shabby, because I see to the minutest detail; and particular, because I won't have work half-done or pay for a whole day where an hour might be wanting.... [W]hen I wanted an Assay Office and room for preparing photo chemicals added on, ... I paid off my Shovel man and made each Shift run out its own dirt after their ordinary time was up, for the same pay, and one of the men do the sharpening and track laying, after that again, for half a dollar per day additional. By this means, without increasing the cost or impeding the progress, I was able to add an Assay Office and Laboratory to the Company's building, which would have cost the Company upwards of £80 Sterling. But by this labor-exacting policy, I have the anathema of the mining population....
>
> As God is my witness, Syme, I am neither working any mine of my own, nor of any other man's—I am not interested either directly or indirectly in any mine except the Imperial. I never employed a man of the Company's one minute, prospecting for me, or in any way on my own business.... My whole time is theirs, and I am never idle—assaying, surveying, preparing photographic chemicals and dark room, examining ground, in [the] tunnel, or reading up on Mining matters. No man was ever more devoted to his work than I am. I am never away from home, except on [the] Company's business, looking for men, getting supplies, attending Meetings where [the] Company's interests require attention; never slept from home since I came here, never had a day to myself except one which [was] employed fishing, and never tried it again.... I know every Car load each Shift puts out, I know how many feet of hole each Shift drills, and therefore, making allowance for the nature of the rock drilled in, I know who works and who plays."[11]

Writing to his brother John, Chalmers similarly poured out his frustration:

> I have got the hardest name out here of any man in Alpine for being a strict and rigorous taskmaster.... If ever a man had his heart and Soul in his work, I am that man.... My house fronts the Tunnel, and my reputation has suffered here ... from my lamp being seen and [myself in] my night dress at the window at four in the morning, frequently watching my night shift that they don't leave before their time....[12]

Intent on confronting the disagreeable rumors at their source, Chalmers wrote to Frank Jones, inquiring pointedly whether he had been the source of the allegations. "I have felt these calumnies very keenly," Chalmers fumed to his brother, "and I shall make him eat dirt before I have done with him."[13]

Jones had indeed been the culprit, though of course he denied it. And behind Chalmers's back, Jones came up with yet another attack: a proposal to "compel foreign miners to pay a license [fee] of $4 per month ... no doubt intended for the 'benefit' of our Company."[14] Luckily, this salvo apparently did not go anywhere.

Despite all the recriminations and drama, work in the tunnel itself ground slowly forward. In his annual report to the Imperial directors dated February 1, 1869, Chalmers confirmed that their tunnel had made it 681 feet into the mountain, and optimistically predicted: "I think I am safe to say that we shall reach the Triumph about the middle of April."[15]

But on the company's financial prospects, Chalmers was less sanguine, at least in letters to his family. "Imperial looks so bad at home (not here) that I am somewhat shaken in my resolution [to send Lewie to mining school]," he wrote his brother John.[16] Sister-in-law Eliza, who'd been caring for his children, was in financial distress, "being frightfully dunned" and "threatened with all manner of disagreeables."[17] With the Imperial's finances becoming increasingly rocky, one director was again floating a plan to rescue the company through bankruptcy.[18]

Winter brought its own unpleasantness. Deep snow had arrived, and living conditions inside Chalmers's house at Bulliona were primitive.[19] "Thermometer two degrees below freezing [indoors] and hands so cold, can hardly write," he wrote to Eliza in mid-January 1869.[20] "[T]he thermometer is always under zero [Celsius] in my bedroom in the morning," he similarly complained to Syme, promising to send him photos "the first day everything is not frozen."[21] To his children, however, he seemed determined to put on a happier face. "The water is frozen in my pitcher every morning," he admitted to son Lewie. "But you know I am not fond of heat."[22]

With Imperial's finances out of his control, Chalmers simply put his head down and did his best to muddle through. He sent reports home describing the minerals that had been appearing in the tunnel—geodes, crystallized calcspar, and streaks of obsidian, although sadly, no silver ore had yet put in an appearance.[23]

Eager to prove his resourcefulness as a manger, Chalmers did not limit his efforts to the tunnel. He forwarded photographs and reports on other mining opportunities as far away as Grass Valley, California, for the company's consideration.[24] He managed to get a new post office established at Bulliona with his own foreman named as postmaster, a step he assured London would "improve the present very unsatisfactory postal management."[25] He had an "excellent" new blacksmith's shop built on the dump below the tunnel mouth, a significant improvement over its original perch above the tunnel, awkwardly accessed by a ladder from the road.[26] Responding to a public complaint, Chalmers also had a "parapet" built to widen the public roadway at the tunnel's mouth. "[T]he sudden issue of the Car from the Tunnel" had often frightened oncoming horses, he explained, causing wagons and their passengers "to sidle towards the precipice" and "narrowly escape a watery grave."[27]

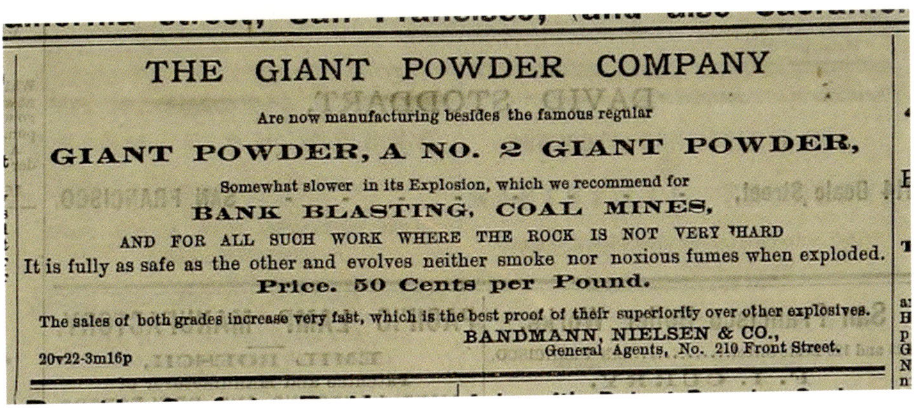

Giant Powder was an early brand of dynamite. The San Francisco company was formed in 1867 to produce the blasting product under an exclusive U.S. license from patent-holder Alfred Nobel. Two early grades of the product were made for blasting different types of rock, both of which supposedly produced no noxious fumes. (*Advertisement from Mining & Scientific Press, 18 January 1873*)

Chalmers also continued his experiments with the new Giant Powder explosive, reporting at length on his observations:

> When I first introduced it, I had a hard battle to fight against the prejudices of the miners, who look upon any deviation from the beaten path as the act of a fool or a madman. My present squad are intelligent and energetic fellows, and were not slow to discover the value of it in a wet hole. Not believing what I was told about its injurious effects on the health of the men, but attributing the outcry to the distaste of innovation, I got two of my men to use it entirely through a Shift of eight hours, during most of which I was in the Tunnel myself. And although while in the Tunnel I experienced no uncomfortable sensations, the moment I reached the open air I was attacked with a crushing headache, but which left me in a few hours. Both the men however, longer in the Tunnel than myself, were seriously affected during the night with headache, sickness and vomiting.... The agents, however, assure me that after a few days, men get accustomed to it, and prefer it to common powder. From experiments I have made with it myself, I am thoroughly convinced of its tremendous power and its instantaneously explosive force. But until the makers are able to prepare it at a cheaper rate and so as to prevent the evolution of the noxious gas which accompanies its discharge, I do not think it will supersede the ordinary blasting powder.[28]

Despite Giant Powder's high cost, he would soon come to prefer it over common blasting powder. The trick was to "get men accustomed to it," he counseled a fellow mine superintendent, noting that supplier Bandmann & Nielsen could provide experienced workers. Otherwise, he admonished, "you will have trouble and, it may be, accidents for some time.... A headache will be made the excuse for poor work, &c."[29]

As March rolled around, Chalmers was again having difficulty meeting payroll, with the "last remittance ... not to hand till after Pay day," and his men were becoming restive.[30] "I was obliged to discharge two men the other day, since which my foreman has had all his hogs poisoned but one, and all manner of threats are made against myself," he noted. Frank Jones, too, was again on the warpath, "vowing vengeance" against Chalmers for relocating several of his claims adjoining the Imperial.[31] Back in London, James Jones was still "trying to get me out of the Managership here, to get his Brother [Frank] in by telling lies," Chalmers confided to his daughter, adding with confidence: "but I have shewn them both off in their true colors. They are very bad men."[32]

By April, the tunnel stretched 790 feet long but was "still in the feldspathic lava."[33] Chalmers acknowledged that he "may have 100 feet to run yet," but boasted he had finally "brought my Tunnel under the perpendicular of the ... Triumph croppings"—his original goal.[34] Still ahead lay the Esmeralda lode, which Chalmers confidently predicted would prove to be a southern continuation of the fabulous Tarshish lode.[35]

Anxious for a fresh set of eyes on their mining operation, the Board sent investor A.F. Wheeler out for a visit in May 1869. Chalmers spent one whole day showing Wheeler over the Company's ground, pointing out "all its beauties," assaying samples from the croppings for Wheeler's benefit, and escorting him to Monitor to witness for himself the prosperous Tarshish and Morning Star mines. "I am satisfied he will write to [Lord Poulett] as he has written to the Board, in the most favorable terms," Chalmers confided afterwards to Syme.[36]

Indeed, Wheeler glowingly confirmed not only the Imperial's potential value but also Chalmers's energy and integrity.[37] This positive report evidently loosened London's purse strings; almost immediately, the company deposited $2,815 in gold to Chalmers's San Francisco account, and another £120 arrived in mid-June.[38]

By early July, with the tunnel approaching 980 feet in length, the miners struck rock carrying sulphurets—the first time this auspicious sign had appeared, and a hint they might be approaching the long-awaited Triumph.[39] But as the weeks passed, no further sign of silver ore emerged.

The longer the tunnel grew, the more necessary good ventilation became. Still unwilling to spend money on a blower, Chalmers had pressed into service an old, 'nearly done' saloon stove at the mouth of the tunnel. Hot air from the stove rushed up the tall wooden chimney, helping to draw bad air through the vent pipes, and causing fresh air to rush into the tunnel.[40] Yet in August, a near-disaster proved just what a poor choice of materials that had been.

> [A]t 2 p.m. Saturday, the Air Chimney at the mouth of the Tunnel took fire and in a few minutes was burnt to the ground. For some days previous, the men had been suffering severely from the effects of foul air emanating from the face, to mitigate which the Stove had to be kept as hot as possible and, being overheated, the wooden Chimney ... took fire at the point of [the] junction [with the ventilation pipe], and was speedily consumed.... Fortunately I was just in time to run across [the footbridge] and get the men out before the Tunnel got full of Smoke. Had my office not been situated where it is, the three men in the Tunnel would, in all probability, have been Suffocated before aid could have reached them.[41]

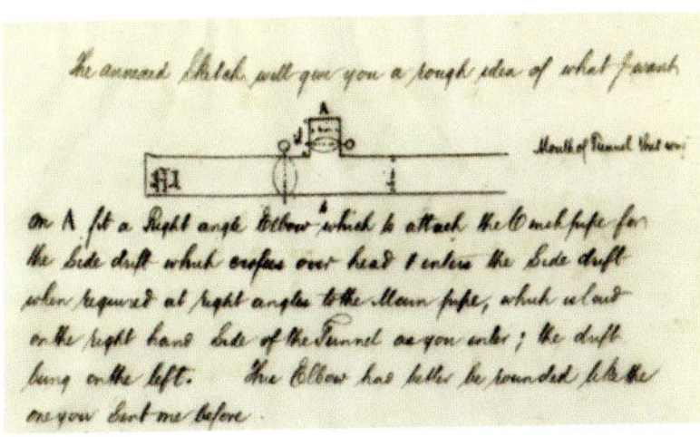
The annexed Sketch will give you a rough idea of what I want

On A fit a Right angle Elbow which to attach the C mal pipe for the Side drift which crosses over head & enters the Side drift when required at right angles to the Main pipe, which is laid on the right hand side of the Tunnel as you enter; the drift being on the left. The Elbow had better be rounded like the one you sent me before.

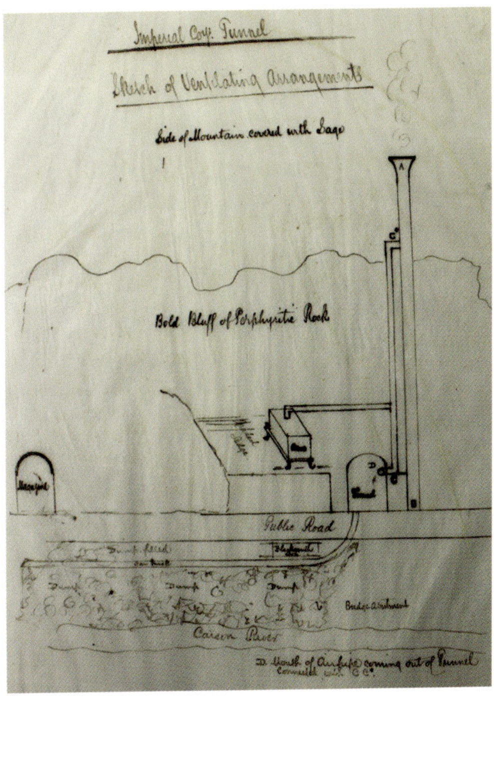

Chalmers shared these sketches of the Imperial's ventilation arrangement with the Imperial directors back in London, including a damper of his own design. Larger sketch shows the arched tunnel opening and smokestack at right, with stove situated on an adjacent ledge to draw air from the tunnel. Powder magazine is at far left. The public road parallels the Carson River, with the blacksmith's shop and dump below the road. (*Chalmers Letters collection, 1867–1870 Imperial Officials, p. 336 and 457 (Alpine County Historical Society and Museum)*)

The Screws Tighten

The Imperial's financial condition quickly grew precarious again. In September, Chalmers wrote not once but twice to remind London that "money is required before next Pay day," October 1.[42]

Though the eagerly-anticipated Triumph lode had failed to materialize, Chalmers was predicting that he had only about a few hundred feet to go to reach the Florence, "when all will be well." He also offered a tantalizing bit of anecdotal encouragement:

> I met a man the other day, whom I think I can trust, who told me that in 1863 he himself took out 500 pounds of ore from the Florence, accompanied it to Sacramento, saw it worked; and the result was $1,200 dollars per ton.[43]

The tunnel now had water gushing from the face, forcing the men to "work in water to their ankles, and in a perpetual shower from above." Still, "at no period since we commenced operations has the face of the Tunnel looked so indicative of close proximity to a ledge," Chalmers enthused.[44] Cautioning that it would be "the height of madness" to stop work in the Imperial now, "not knowing but the next drill hole may tap the treasure we have worked so hard to reach," Chalmers urged the company to either send him money or cable him to "stop."[45] They did neither. The last of October came and went without funds for pay day.

Now overdrawn by $78 at his bank and in need of winter supplies, Chalmers once again admonished London: "If money does not come by next Pay day, I must Stop work."[46] The threat finally worked. A remittance of $1,391 in gold arrived mid-November, providing the immediate relief needed. Still, Chalmers cautioned, "Money received but won't carry me to the 1st [of] January ... [Y]ou can fancy my position here among a set of roughs and no money."[47]

Bankruptcy was again being talked about for Imperial, and Chalmers expressed himself all in favor. He'd been urging reconstruction "ever since I came here and found how things were," he told Syme, adding that "it must be done to save the company from wreck."[48]

Chalmers's own personal bankruptcy still had not been concluded, either. He penned a disgusted letter to his trustee in Aberdeen, emphasizing that he had "made a full and exhaustive surrender [and] left Scotland as a pauper," and asking "what prospect there is of getting my discharge." Without such relief, he swore, "I shall never return to Britain."[49]

His family, too, was suffering from lack of funds. Chalmers gratefully acknowledged the kind intervention of Lord Poulett and another subscriber, whose help had "relieved my family of a very unwelcome visitor—the Landlords' Bailiff."[50]

Despite his rather desperate financial condition, Chalmers took advantage of a team going to Reno in December to order $400 dollars' worth of powder, candles, and track, so the work wouldn't be "brought to a standstill in the depth of Winter."[51] With the tunnel now reaching 1,292 feet, he urgently reminded Syme:

> On 1st [January], without a Cent for myself, I shall require $1,181.61 ... By 1st February I will require $624, 1st March $624, 1st April $624, [totaling in all] $3,453.61, of which

> [I require] $2,605.61 on receipt.... It is all very well for you to sit at 114 Palmerston Buildings and with your 'Mark Me's', ask me first to live on 0 0 0s, and then to run the Tunnel without money, in an outlandish place like this. But it is easier to advise than practice; and were you here, you would feel it so. No tick here, no Convenient Uncle, no faith in promises ... [H]ow the Dickens I am to carry on the work till 1st February even, I don't know. Send me what money you can and rely on me doing my best.[52]

Poor as Chalmers was, he was determined to indulge in a favorite hobby: "turning," the art of creating ornamental spindles. In mid-December, he sent off for several catalogues, looking for lathes "strong enough to turn iron or brass" and equipped "with screw cutter and slide rest."[53]

More disturbing news arrived at Christmas, with Eliza writing to report that Lewie's doctor had issued a poor prognosis. "I don't give a fig for Dr. Brown's opinion," Chalmers retorted. "[If] Lewie comes to me, mark me, ... I will have him ... well and strong in six months. I believe Ealing decidedly unhealthy, and Brown a humbug."[54] Chalmers pressed Eliza to relocate with the children to a "cottage away in the country," or perhaps find "some old chateau" in France. He also penned worried letters to Lord Poulett and Syme, asking Syme to send Lewie out to Alpine in the company of Chalmers's brother, John, and hoping Poulett would help secure a naval nomination for his younger son, Willie.[55]

As for young Alick, now fifteen, Chalmers wanted him to finish his studies in London, then go on to Oxford, but Alick was expressing a sudden interest in seminary studies. "As to [Alick's] being a clergyman, that will be all in time enough," Chalmers wrote Eliza, adding: "Generally I have a thorough contempt for the cloth, and my acquaintance with individual members of the Set has not improved my opinion of them in their corporate capacity."[56]

On the positive side, subscriptions were starting to come in for the Exchequer Gold & Silver Mining Company, newly formed to work the Buckeye claim. Brother John

Although Chalmers claimed shortly after arrival that he had given up both tobacco and "grog" as a cost-cutting measure, he would later order pipe tobacco, cigars, and kegs of good Scotch whiskey as part of his supplies. (*Harper's 1879; Author's collection*)

telegraphed encouragingly: "Will be a great success soon." Earl Poulett, the earl of Buchan, and the Lord Louth had all agreed to serve as directors for the new company, and the earl of Carthness and the duke of Sutherland were also expected to participate.[57]

"I have just succeeded in floating another Company in London—The Exchequer," Chalmers enthused to fellow entrepreneur, George H. Ensign. "Good Silver Company, Capital $300,000 to work Buckeye."[58]

Impressed with specimens taken from the Scandinavian Canyon mines, Chalmers hoped to send home "a few miniature silver bricks, as a foretaste of what they may expect hereafter."[59] "My boys are all excited at the present moment with the appearance of the [Buckeye] Tunnel and the immediate prospects of a ledge," he wrote Syme in late December, adding "Keep this to yourself."[60] His plan for this new mine was to run the upper Buckeye tunnel during the winter, stockpiling ore that could be processed in the spring to give him "some bricks to send home at any Cost. This will give confidence and do us all good."[61]

Despite the winter weather, Chalmers had already commenced "laying off [surveying] a Road to reach [the nearby] Mill before the Snow got too deep," finding that "a good road can be made with an easy grade for a reasonable sum, by which we shall save buying the toll Road and [also] some three miles of teaming."[62] He was also arranging housing for his men near the Buckeye tunnel, writing to the trustees of the I.X.L. for permission to use their existing buildings "during the Winter Months." While he was at it, he casually inquired, "I have heard it rumoured that the I.X.L. is for sale. If so, on what terms."[63] He dispatched a similar letter to a representative of the Lady Franklin mine (another lode in Scandinavian Canyon), asking if his men could occupy "the Franklin House till Spring," and hinting that "Mr. Davidson might do a worse thing than hand over Franklin to us on my terms."[64]

Chalmers told Syme that he hoped the Exchequer company would eventually spend the money to erect a Stetefeldt furnace, asking him to "ascertain at what price we can secure from Stetefeldt the right to use his furnace, now in successful operation at the Aurora Mills, Reno.... I am satisfied," he added, that "this is the process for the economical and exhaustive beneficiation of ... all our so-called rebellious ores, rich and poor."[65]

With fresh hope now filling his sails, Chalmers had his eye out for other investment opportunities as well. He dashed off a letter to Ensign, inquiring, "On what terms will you sell Morning Star?" He also asked about Ensign's Lake Tahoe water project, assuring him that he had a "Scotch Millionaire friend in London who is open to such a Speculation, on a good Shewing."[66]

Death Knell for the Imperial, On to Exchequer

By the end of 1869, the Imperial tunnel had reached 1,338 feet. Although no silver ore had yet appeared, it was in what Chalmers called "very encouraging indications."[67] Those included a small ledge "filled with sulphurets of both iron and copper," which, when assayed, showed "distinct traces of gold."[68]

But the Imperial's finances were rockier than ever. A letter from William Ralston of the Bank of California in January 1870 advised Chalmers tartly of the "state of his account." Chalmers frankly admitted overdrawing his balance by $729, but protested that he would not have done so "had I not expected immediate remittances." He blithely assured Ralston that "the Imperial Company is being reconstructed, in order to get rid of the enormous vendor's Capital, and ... it will take three months to do so, when all will be again Serene."[69]

On January 31, 1870, Chalmers wrote Syme to advise that he had made another 16 feet in the tunnel the previous week. "But you mustn't expect good running for no pay. I want to pay off a lazy devil of a Carman, but he knows I can't. Pleasant—very," Chalmers added caustically.[70]

Even so, his faith in the project remained strong. "Two hundred seventy-six feet more will put us all in clover if they can only raise money enough to pay off our present liabilities (some $2,100), and carry us on to the 1,636," he wrote Wheeler.[71]

But creditors were pressing. Chalmers's overdraft at Bank of California had mushroomed to $926.03, and he had been forced to borrow an additional $260 to "pay telegrams, cutting wood for Imperial &ca &ca." Holders of the dishonored checks were harassing him personally. "I was attacked yesterday by the holder of one for $89 odds," he reported, adding, "I am now without a Cent to buy necessaries, or to pay my Servant. I have flour in the house, on which I could get along, but I don't like my Servant telling about that we have no butcher meat." From his $1,140 salary over the past year, he had kept just $480, after expenses, Chalmers complained, "to live and dress on in a Country where everything is unexceptionally dear. Verily I am a lucky dog.... Now Syme, my dear fellow, send me this money [$2,295 for December pay, suppliers, dishonored checks and cash borrowed]. I really cannot face the storm without it."[72]

Reconstruction of the Imperial had finally gotten under way, and Syme kept promising that additional funds were coming. Ever hopeful, Chalmers assured his men of the same. But on February 5, 1870, he was forced to officially acknowledge receipt of the dreaded order: "Stop work."[73] It was the death knell for the Imperial Gold & Silver Quarries, Ltd.

By the time the work was halted, the Imperial's double-track tunnel had been pushed an impressive 1,406 feet into Mount America.[74] Chalmers would later brag that the Imperial "was the quickest run Tunnel in this Country of its size without drilling machinery."[75] The tunnel, he claimed, had managed to cut ten ledges through "hard porphyry," with the work only being suspended "until the success of [the Exchequer] shall give fresh encouragement."[76]

Lawyer that he was, Chalmers promptly ensured that mechanics' liens were filed against the Imperial's assets for both his miners' salaries and his own.[77] He also had one small ace in the hole: a mining location he had quietly filed three years earlier on 10,000 feet of "the same ledges as Imperial and immediately adjoining on the South," from which he intended to "make a pile," he told Syme, advising him to "keep mum" and promising "one-half to you of course ... I thought it useless to speak of it till Imperial succeeded."[78]

On February 14, 1870, just nine days after Chalmers had been forced to stop work on the Imperial, his younger brother, John, showed up on his doorstep in Bulliona.[79] It was horrible timing, but John, with even less mining experience than his older brother, had been sent out to serve as the nominal manager of the newly-formed Exchequer.

Despite the Imperial's troubles, work on the Exchequer had been moving merrily forward for the past few months. Having successfully secured use of the I.X.L. and

Franklin boarding houses for the winter, Chalmers had already hired workmen and set them to cleaning out the old Buckeye tunnel, "in which there has been very considerable caving." They had fitted up a blacksmith's shop and begun laying new track. The men were currently sinking an incline from the tunnel along what appeared to be a good seam, in hopes of getting out enough rich ore over the winter to produce bullion in the spring. Ore samples sent to Johnson & Matthey had assayed even better than Chalmers's own results.[80] Chalmers had also staked a claim to a "North Extension" of the Buckeye Lode, "an acquisition which you may safely regard as almost doubling the value of your property."[81]

With his right to occupy Imperial House no longer a sure thing, and a brother to share his lodgings, Chalmers pitched the Exchequer Company on a fresh acquisition. "I now come to what I consider a very essential element to the Success of your undertaking; I refer to the purchase of the Davidson Mill," he wrote. Amazingly, the mill's owner was willing take half the purchase price in shares, and the only thing lacking was "a Stetefeldt Furnace for cheaper roasting; although it has the ordinary Roasting appliance [already]. With a few Repairs it can be in working order tomorrow," he assured the Exchequer's directors. A sawmill and water privilege would come with the deal, along with buildings to house both the quartz and sawmill employees. And last, but hardly least: "Your Manager must have a house, [and] there is one ready for him.[82]

The two-story house at Davidson's Mill was a far more substantial dwelling than the hastily-built Imperial House. And though it would take another year to accomplish, Chalmers's heart must have swelled a bit at the thought of improving his circumstances by moving there. Little did he know the pain and pleasure he would experience within those walls in the next fifteen years.

Davidson's two-story house would soon become known as "Chalmers Mansion," the name it still carries today. Although the caption to this image said that the view was "from the North," it would more accurately be the northeast; the view is looking toward the southwest. (*Huntington Library, Withington stereoview, ca. 1876*)

4

Launching the Exchequer (1870–1872)

It must have been a bittersweet moment for Lewis Chalmers when his younger brother arrived in Bulliona, hard on the heels of the Imperial's demise. John had been sent out to serve as the manager for the new Exchequer Company. Lewis, although already here, had been skipped over for the post. "My name stinks," he wryly admitted to Syme.[1]

John's appearance on February 14, 1870, did bring with it some good news. For one thing, dispatching John all the way across the ocean confirmed that the Exchequer had been successfully floated, with some £3,000 raised in subscriptions. John also brought with him a welcome infusion of cash for the new company: four "bank post bills" of £100 each, the equivalent of $1,954 at the time.[2]

Perhaps most important of all, John's title as manager came with a promised salary of £400 per annum. Having helped his brother to secure that appointment, Lewis was expecting to share the income—at least until the Imperial could be resurrected, when his own salary might resume. The modest salary was hardly enough to meet Lewis's family obligations, however, and "the State of things at home is not agreeable," he confided to friend Arthur Wheeler.[3] But it was better than nothing.

Now with two solid years of mining experience under his belt, Chalmers was confident in his own abilities. Yet John was entirely green about mining affairs. The real work of managing the Exchequer, they both knew, would ultimately fall to Lewis. "The Imperial [John] could manage with a few instructions," Chalmers confided to Syme, "but sinking, judging ore, shaft timbering, knowing when men do a fair day's work, is not learnt in a day."[4]

Although work in the Imperial Tunnel had been halted, Chalmers declared his intention to continue to occupy the company house at Bulliona "to look after the Company's property generally."[5] He had already added a wall to his bedroom in anticipation of John's arrival, splitting the room in two to give John a space of his own. But with no money to continue to pay his housekeeper, Lewis lamented, he would soon be forced to "turn cook, and housemaid and laundress."[6]

John took up his new duties promptly after arriving, going over the Exchequer's books and trekking through the snow to Scandinavian Canyon to examine the Buckeye No. 2 mine in person. Just two days after setting foot in Bulliona, John

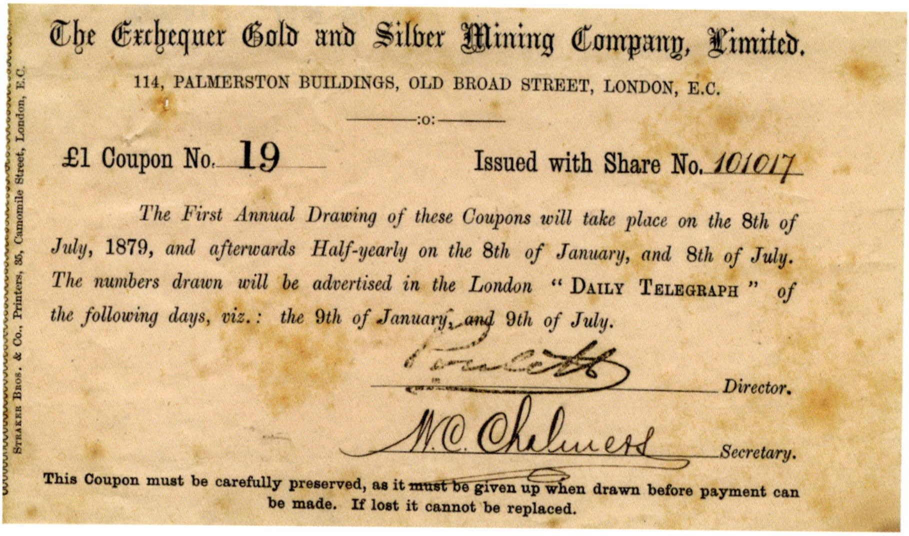

The Exchequer Gold and Silver Mining Company, Limited.

114, PALMERSTON BUILDINGS, OLD BROAD STREET, LONDON, E.C.

:o:

£1 Coupon No. 19 Issued with Share No. 101017

The First Annual Drawing of these Coupons will take place on the 8th of July, 1879, and afterwards Half-yearly on the 8th of January, and 8th of July. The numbers drawn will be advertised in the London " DAILY TELEGRAPH " *of the following days, viz.: the 9th of January, and 9th of July.*

Poulett *Director.*

W. C. Chalmers *Secretary.*

This Coupon must be carefully preserved, as it must be given up when drawn before payment can be made. If lost it cannot be replaced.

STRAKER BROS. & CO., Printers, 35, Camomile Street, London, E.C.

Coupon entitling a shareholder to payment by The Exchequer Gold and Silver Mining Company, with signatures by Lord Poulett and William C. Chalmers, Lewis's son. (*Dustman collection*)

wrote to the directors in words that sound suspiciously like his brother dictated them. He confirmed that he had found the accounts "satisfactory" and that work at the mine was "being prosecuted most effectively." The Buckeye Tunnel, he reported, was 6.5 × 5 feet in size and already over 350 feet long. Crews were in the process of sinking a 6 × 4-foot incline at 147.5 feet from the tunnel's mouth, and a chamber had already been excavated above the incline for a windlass and hoist.[7]

Visiting a working mine in the middle of winter proved a strange adventure for the forty-year-old John. He penned a humorous note to Syme on the same day as his prim-and-proper official report, offering a more candid reaction:

> I have just returned from the Mines after floundering about in snow drifts for about an hour. If there is any faith to be put in Mines, it appears to me that we have at last hit upon the right thing in Buckeye No. 2. The only drawback to mining exploration, I find, is having [to] swing about in the air by a slender rope and having one's nut knocked to pieces on the sides of the shaft in getting up and down.[8]

There really was rich ore in the Buckeye No. 2 and other Scandinavian Canyon lodes, though deposits were spotty and often found only close to the surface. But there was no way at the moment for Lewis Chalmers and his brother to know that. Publicly, Lewis declared himself brimming with confidence about the new mine's potential. In a letter to a potential investor from Whitehall Gardens, London, for example, Chalmers bragged that his assays of Buckeye ore had confirmed results as high as $1,232 per ton. "[I]t is exceptional to find rich ruby ore at the shallow depths it has been cut," he enthused, adding confidently: "the received opinion is that silver ores improve with depth" and urging "[n]ow is the time to buy."

> I have pinned my reputation to this Mine, and from all appearances it is not likely to play me false. My foreman has just left me with the remark, "I have seen a good deal of ore in my time, but I have never seen anything like this...." [W]ith proper capital to commence with [this will be] the Mine of this Section.[9]

Privately, however, Chalmers confessed to worrying about Exchequer's financial stability. Although the mine itself was "[i]ntrinsically ... a first-class property," he told Syme flatly that predictions of a grand success might be premature: "[I] don't see it; the same beggarly, hand-to-mouth thing as the Imperial, *qua* Capital."[10]

Nevertheless, Chalmers understood all too well that *seeing* drove believing. The key to ensuring both happy investors and a continuing flow of working capital was hard evidence of actual silver, and to that he turned all his energies.

A Buckeye Brick

Determined to produce a small silver brick from Exchequer ore, he demanded that his foreman focus on extracting a few tons of cherry-picked ruby ore from the incline, and "[l]et everything else give place to this." Sure enough, by March 1870, the crew had 3.5 tons of ore ready to ship to the Auburn Mill at Reno. Chalmers planned to personally observe its working, "that we should be acquainted with all the details of the 'Stetefeldt' process."[11]

To prepare it for shipment, the ore was packed into sacks of one hundred pounds each, and the mouth of each bag was carefully sewn shut to guard against loss or tampering. But "much to our disappointment," snow threw a monkey wrench into Chalmers's plans. On March 5, 1870, just as the wagons were about to depart for Reno, the teamster refused to accept more than 2.5 tons due to the poor condition of the roads.[12] As if that were not disappointment enough, before the pared-down shipment could even reach the Reno mill, Chalmers received word that the mill had been forced to shut down unexpectedly for repairs.[13]

Despite such obstacles, the scheme eventually worked. On April 9, the Exchequer forwarded its first silver bullion to London, a small brick of purity .901 fine.[14] The overall yield at the mill had amounted to only $140 per ton, though Chalmers declared himself well pleased with this result, "[c]onsidering that the ore was not sorted."[15]

Tangible proof of the mine's potential in the form of that silver brick gave Chalmers bragging rights indeed, and he wasted no time putting his own silver tongue to work to promote it. "[N]othing better could be wished than is to be seen in the upper tunnel on the Buckeye ledge," he pronounced in a report for a mining publication. Soon, he predicted, the Exchequer would keep the Davidson mill "in full blast," hinting that "selected specimens of the best ore assay as high as $2,000 per ton, and all the ore contains more or less gold."[16]

John Chalmers, too, puffed the mine's prospects in his official report to the Exchequer directors, observing that "the schutes of Ore are not only getting more frequent, but improve in quality as we go down. Patches of clay, even, have assayed as high as $487.62 per ton."[17] For a few still-dubious directors back in London, Lewis admonished:

> I only wish I could squeeze the whole lot of your unbelieving Directors for five minutes into the bottom of our Winze (South drift), and if funds weren't forthcoming immediately thereafter, why, you don't deserve to own a good Mine."[18]

In addition to the Buckeye's recent success, however, it also had run into unexpected hurdles. "On reaching the 354 [foot mark], the Air got so bad that the men could scarcely work, a contingency at that distance not anticipated," John was forced to report. A "large bellows and hose" were pressed into service for a temporary fix. But a more permanent ventilation solution was required. Obtaining milled lumber to build air boxes was an "utter impossibility." So instead, they were forced to adopt a far more labor-intensive answer: crews were set to cutting a 100-foot air shaft to reach the surface.[19]

Alas, Poor Jones

Meanwhile, much to Chalmers's delight, he learned that his enemies, the Jones brothers, had been caught up in troubles of their own. Thanks to a complaint by Oscar Thornton, James Jones was arrested and thrown in jail at Monitor. He was given the choice of continuing to cool his heels behind bars at Thornton's expense, or being released to a "keeper," to whom he'd be required to pay board. Jones unsurprisingly chose the "keeper" option, and was released to the custody of Undersheriff Michael Maxwell.[20] But he would not stay with Maxwell for long.

After telling friends they planned to visit the Leviathan, both Jones brothers "decamped on foot over the mountains"—much to the chagrin of Frank's wife, who was left behind.[21] "There is a judgment in this County against [Jones] for $4,000 and the Michigan boys are looking [for] him," Chalmers gleefully told Syme.[22]

Chalmers also expressed satisfaction at having out-foxed the Joneses by acquiring the Sonoma claim, a northern extension of the Buckeye No. 2. The Joneses had attempted to purchase it for unpaid taxes, but Chalmers managed to enlist a former shareholder to redeem the mine for him from the tax sale. "[T]he property is gone from Jones," he gloated. "Alas poor Jones. Leave L[ewis] C[halmers] alone next time, Master Jones. You'll surely tire of playing a losing game always."[23]

Now, Chalmers was eyeing yet another claim, the Fremont, which adjoined the Buckeye on the south. Not only did the Buckeye ore seem to be running in that direction, he assured Syme, but the Fremont ground was the perfect place to erect permanent hoisting works. The additional ore they could expect from this new claim "likely will pay [for] your Mill 10,000 times over," he predicted, urging that the Board "must have [it] on any terms." Were the Fremont's owners to get wind of Buckeye's success, he cautioned, "ten times the price asked won't buy it."[24]

> [T]he Schutes of ore (yesterday's assay close on $1,400) are at present all trending South into that ground [Fremont], and the Miners are getting so excited they will blab and brag, until the rich discoveries reach the wide-awake General Hall, who is Yankee enough then to know how to make his own terms.[25]

Scheming, Dreaming

Chalmers continued to stress the necessity of a mill, again pushing London to acquire the Davidson Mill on Silver Creek. At three cents per pound for freighting to Reno, carriage alone would eat up $60 for every ton of ore, he pointed out, not to mention the costs of milling.[26] Davidson's property also came with a handy timber ranch and a water privilege.

Simply purchasing the mill, however, was not the end of his wish list. The Davidson Mill featured an old-fashioned sixteen-stamp mill for crushing, but Chalmers proposed improving it with an eight-stamp steam mill and a Stetefeldt furnace, at an estimated total cost of between $35,000 and $46,000.[27]

Chalmers's first attempt to craft a deal with Davidson quickly fell apart. Davidson was demanding $20,000, with half of that sum paid almost immediately and the balance by year-end. When Chalmers assured him that proposal was a non-starter, Davidson wrote back to say he had other buyers in mind and "the Mill was no longer for sale."

Chalmers was convinced, however, that this was simply a *ruse de guerre* to seek better terms. "He is still poor," Chalmers assured Syme, pressing the firm to make another offer. "[M]aybe by the time I hear from you [he may] begin to find himself wrong in his expectations from other sources."[28]

Nothing fell into place immediately with the proposed acquisition of either the Fremont claim or the Davidson mill, but Chalmers's fears about being forced to become his own cook and maid had at least proven groundless. In July 1870, the census-taker found Lewis and his brother John still living in the Imperial house at Bulliona, along with a housekeeper named Annie Kelly.

Turning his attentions back to the Imperial, Chalmers now pressed the stricken company for money to purchase a Leschot steam-driven core drill to allow them to speedily prospect their host of ledges. This new invention would not come cheap; the estimated price tag was $3,750—the equivalent of nearly $80,000 today. Yet instead of the four months it would take to sink 200 feet using picks and hand-drilling, the Leschot was said to be capable of boring 100 feet in one day of two shifts. In just a year's time, Chalmers figured, the Leschot could prospect fifty-two times the distance possible with old-fashioned tunneling and would have "paid for itself."[29]

> [Y]ou have eight other ledges [besides the first one] ... which ought to be prospected otherwise than by your Main Tunnel, as years must elapse before some of them can be reached in this way, and you want to know at as small an expense and as speedily as possible whether those ledges, which show so well on top, are worth anything at a depth of from two hundred to six hundred feet, before working them in the ordinary way. I think the Leschot Drill will solve the problem.... Professor Rodolphe Leschot, a Civil engineer of Paris, found by practical experiments that a rotating drill, armed with diamond points placed round the periphery of a disc, could be made to bore holes in rock to great depths and with great rapidity, by forcibly injecting a Stream of water into the bore hole through the interior of a hollow drill.... By multiplying these bore holes, you may even approximate the quantity as well as the quality of the ore, and you can also ascertain the presence or absence of water, and the dip or angle of your ledge.[30]

Not surprisingly, the Imperial's London directors did not leap at Chalmers's latest suggestion. The tunnel remained closed as efforts to restructure the company's finances dragged along. "I am anxious to see the Imperial revive," Chalmers declared, "were it only (selfish, perhaps) to prove that I have been right in my appreciation of it."[31]

It was a long, slow process, but the $6,987 in mechanics' liens that Chalmers had filed on behalf of himself and his workmen at least had the benefit of ensuring that no one else was able to take over the tunnel, blacksmith shop, outbuildings, and Imperial House in the meantime, without fronting that hefty sum.[32] In the coming months, Chalmers not only perfected those liens by filing suit but also obtained a formal judgment against the Imperial.[33]

As spring 1871 arrived, Chalmers was still scheming to pick up other mines. He had enticed the owners of the I.X.L. to part with their stock on attractive terms, and now was "after Buckeye No. 1." As if that was not enough, he could also brag that his workmen had just hit a "beautifully defined ledge" in Buckeye No. 2, yielding $1,900 ore.[34]

Money to continue Exchequer's operations was still tight. As a cost-saving step Chalmers cut his windlass men's wages to just $3 per day. Not surprisingly, "[t]hree batches of men (twelve) have left me, and I am now on the outlook for [others]," he wrote in disgust. "Too hard work, they said."[35]

Back in London, Chalmers's detractors continued to grumble. Some were comparing his winze to a water closet [toilet], prompting his feeling retort: "Treat winze as a water closet, and shit on Silver? You forget that here only have we got ore." He staunchly defended his decision to run a small prospecting shaft, speculating that the ledge he was following could turn out to be a bonanza: the "Comstock South."[36]

Chalmers kept up with developments around the world through the mining press, and his letters demonstrate just how knowledgeable about mineral lore he had become. At one point he dashed off a dissertation on jacutinga (a gold-bearing iron ore) and other types of mineral formations found in Brazil.[37] It was that depth of knowledge that allowed Chalmers to sound so certain as he reassured Syme:

> In the very best of mines ... there are the times when, running long and unsuccessfully, you are apt to have surgings of misgiving. But I always went back to the old thing, it is impossible that in such an immense true Ledge as this we have no more such ore as gave the Lode its character.... [The] Comstock produced only in ore one-seventh of its entire length, and that one-seventh did the business.[38]

Likely realizing that Chalmers's persuasive abilities were the best answer to detractors, and that those abilities would have the strongest effect in person, Syme urged him to "come home at once." No, Chalmers responded firmly. He was determined to stay and "get to my ore," though he complained heartily about his current situation.[39]

> I have told you why I will not return. I may lose character, I admit.... What a delightful thing it is to be a poor and abused man with a large family.... Sunday and Saturday, nothing but this cursed mine, night and day, scheming—and [for] what?

> Well, doubt!... [Leaving] lets me leave this d[amne]d thing a failure, notwithstanding all the care I have bestowed on it. I [would] go into the world a mere drudge—in short that I've lost my chance, an admission I never thought I'd have to make.[40]

In a series of personal asides, he confided to Syme that "worry has made me gray!" while, to capitalist J. T. McKenzie, he ruefully admitted: "You once told me I was a fool to have anything to do with [Mines]."[41, 42] Yet he felt too old to do anything else. "I think my day has gone by. The old horse works hard, but to little purpose: [I get] hay, perhaps, but no oats."[43] Saddest of all: "[E]very dog has his day. I've had mine."[44]

Given the time, energy, and reputation he had invested, he felt stuck in Alpine: "I can't see my way very well to get out of it."[45] Yet despite his most diligent efforts, he saw the company's fortunes slipping. As he confessed to Eliza, "I am afraid Exchequer's days are numbered unless certain arrangements now on the tapes are completed."[46]

Adding to his discomfort, Chalmers and his brother were not getting along. Lewis hinted he might quit and go his own way, suggesting that John be put in charge if a new company was eventually formed for the latest claims. "If you float [I.X.L. and Buckeye No. 1], won't you give John the Secretaryship? I can get something else to do, I think possibly, in this Country. I'm afraid he couldn't," he wrote to Syme in May 1871.[47] But Alpine really was not John's cup of tea. A slightly later letter shifted tack: "I want a secretaryship for [John], at home. It would suit him best."[48]

Chalmers's family situation at home was also deteriorating. Eliza Chalmers, the sister-in-law caring for his children, was facing bankruptcy. "[H]ow she can owe so much money, £1000, I don't know," he fretted. "Two years ago I advised her to go to the Country, but no."[49]

Also, his teenage sons were misbehaving. Alex was now seventeen, while Willie was fourteen, and both, Chalmers declared, were "very bad boys. The former drinks, swears and smokes; the latter not much better." Among other things, the pair had "amused themselves pitching the furniture from the second-floor window."[50] "Puberty be hanged," Chalmers groused, after Syme had offered that as an excuse. "We all arrived at puberty without making beasts of ourselves."[51]

Chalmers wrote son Alick an impassioned letter, condemning his 'blackguardism,' and asked Syme for recommendations about where his son could be sent off to board with a clergyman or "retired officer's family" to halt his "downward track."[52] Both boys needed "kind and intelligent male supervision," he sighed. But with problems enough already on his plate where he was, Chalmers made clear that sending them out to California was not an option.[53]

It was not that he didn't care; he simply could not stretch in one more direction. But it was another depressing reminder that he was not able to meet his family obligations. To Eliza he wrote, "Sunday and Saturday here [are] toil, toil; sleepless nights, anxious days, and all for what. I can write no more."[54]

The House at Davidson Mill

Davidson at last began to come around about selling his mill. By May 1871, Chalmers confirmed excitedly to Syme, "I expect to be able to send you in ten days an offer of a fine wood ranch, saw mill with 40-hp engine, shingle mill and water power, at $5,000

[down].”[55] As thinking small was, like shabbiness, not in his composition, Chalmers would soon suggest that Exchequer expend $171,000 to erect a thirty-stamp mill and Stetefeldt furnace.[56] Naturally, additional equipment would be needed at the mine, too: “What I want is an engine, boiler, hoisting drum and pumps sufficient to sink 1,000 feet, wire rope, kibbles, car platform, etc. for a shaft five-foot-two by four feet in the clear.”[57]

Selected specimens from the upper tunnel at Buckeye assayed as high as $2,500 per ton—specimens that Chalmers proudly sent home.[58] Though forced to admit he no longer held the same “excited opinion of the Mine” as he had at first, Chalmers added confidently, “I am after what I know can’t be far off—a large body of rich ore.”[59]

Perhaps it was those ore specimens that did the trick. The board finally agreed to Davidson’s terms. It soon became obvious that the Davidson property was not perfect; the cam shaft and foundations at the quartz mill both needed repair. In addition, the worm drive at the sawmill was damaged through an operator’s mistake.[60] But as Chalmers had pointed out earlier, the property came with a lovely two-story house, putting the manager much closer to the mines.

That spring, Chalmers and his brother John moved up to the former Davidson house at Silver Creek, a larger and more substantial residence than they’d enjoyed at Bulliona. About that same time, Chalmers’s son Lewie came out to join them. Lewie was still unwell, but Chalmers was convinced that the dry Alpine climate would help his tubercular lungs.[61] Enough money was also forthcoming from London to enable Chalmers to purchase four horses, a wagon, and a heavy log wagon, all for $1,250.[62]

But by July, the Exchequer’s fortunes were again teetering on the brink of disaster. Chalmers was overdrawn at the bank, his men were unpaid, and he had been forced fire both his foreman and his cook.[63] In high dudgeon, Chalmers complained bitterly to London:

The former Davidson house, now Chalmers Mansion, with “Chalmers Peak” behind, with likely members of his mill crew. (*Huntington Library, Withington stereoview, ca. 1876*)

> I owe Davidson and Mrs. Hanson, my butcher, etc., and have to sell my furniture, bedding and transit instrument to pay them. We can now scarcely get credit for a pound of meat, and I owe for that only about $14 or $15.... We are all going about in mended rags, the worst dressed men in Alpine County—a disgrace to the Company. We live worse than any of the miners.[64]

In case his own representations about the value of the mine had not been enough, Chalmers advised that an expert from Empire City named Coffin had recently paid him a visit: "I showed him specimens of Buckeye (Exchequer); he opened his eyes and said he had never seen finer ore except in the 'Ophir' at its start in Virginia City."[65] The company had already spent $42,355 on its mine and mill, he urged. What a shame it would be to shut down "just as something good was transpiring."[66]

A flurry of excitement swept through the Alpine mining community in late July, when the Globe mine at Monitor struck a vein of rich copper ore—so rich it was said to be 50 percent copper, with silver, too, "said to be in large quantities." Yet unable to get machinery to sink on the ore, the Globe was shut down again "by the men not paid" after just a few short weeks.[67]

By August 1871, Chalmers was three months in arrears paying his own men and having trouble feeding his oxen.[68] Sinking deeper and deeper into debt, he was afraid his father's watch soon would "have to go."[69] "I can't sleep at nights lately," he admitted to Syme, later admonishing: "It would be simple suicide to stop, so near the goal."[70]

Even amid such financial distress, Chalmers kept his eye out for other mines to buy. By borrowing $50 from his housekeeper, he managed to make a trip to Mono County, California in mid-August to view the Dunderberg mine, which

The Globe Mine in Monitor, *circa* 1871. (*Inset from larger panorama; courtesy of the California History Room, California State Library, Sacramento, CA* (*Winchester Collection*))

Dr. Munckton of Carson City had been pressing him to buy. Assays of samples brought back did not live up to Chalmers's expectations, however, and he decided Munckton had "got [his] ideas too high up."[71]

His acquaintance with Munckton, a pharmacist, did prove useful, however, when Lewie fell ill with an intestinal ailment. "Many thanks for the drugs," Chalmers wrote Munckton in September. "The chalk mixture stopped the diarrhea and, though weak, Lewis is much better."[72]

Finally, a welcome telegram arrived. On September 3, Chalmers received word that £200 was on the way for Exchequer—and that the long-shuttered Imperial had been resurrected as the "Sovereign."[73]

The much-needed funds showed up, as promised, a few weeks later. "The other £200 has come, and I will do my best," Chalmers promised, itemizing the work that still lay ahead:

> Flume to repair, salt and quicksilver, hay and barley for oxen, brick and lime for retort to buy, and all these workmen to hire, [all] on $969.70 less overdrafts.... I am now busy with my ore platform and tramway, which will take me ten days or a fortnight of all hands.... I only hope snow will keep away, and I will yet be the first shipper of bullion from Alpine, notwithstanding all the Monitor crowing.[74]

Negotiations to acquire I.X.L. stock for the company were continuing, and Chalmers was hopeful that the annual meeting of I.X.L. shareholders would soon approve the sale.[75]

Yet each step of progress was, it seemed, accompanied by an equal step backward. By the end of September, the workmen were again clamoring for three months' back pay.[76] Worse yet, Lewie's health took a brief turn for the worse in November—an attack of "British Cholera or something similar," Chalmers called it, which "threw him back very much," though the lad soon recovered.[77]

The workings of the old Davidson mill, too, were giving them fits. A short run on ore "by way of experiment" produced a small quantity of bullion an impressive .967 fine, valued at $305.77. But a mortar block quickly gave way, and other parts kept breaking. "No sooner do I get one thing right than another goes wrong," Chalmers chafed, adding in a separate letter: "Bad luck shadows us."[78]

By the time 1871 drew to a close, the old mill was shut down for repairs and the Buckeye mine itself had filled with water. "[T]he reason they have let the Shaft fill up is because winter caught them napping again," crowed laid-off superintendent George Lowe, no fan of Chalmers. "I was not surprised when I heard that they had shut down the Mill, and as long as Chalmers has the management of it, it will never work right."[79]

The fresh new year brought fresh possibilities. Chalmers's friends in London managed to successfully float the I.X.L. Company, and in January 1872, a stunning remittance of over $20,000 arrived to fund the new company's operations. Like the neighboring Buckeye, the I.X.L seemed promising—discovered in 1861, it had been one of the earliest claims in Scandinavian Canyon.[80] Chalmers's own reputation had also been rehabilitated enough that he was officially named manager, a position (and a new salary) that he gratefully accepted. As usual, Chalmers had lots of ideas for spending all that money. He began looking into portable hoisting equipment and investigated Hallidies' Endless Wire Ropeway.[81]

In private, however, Chalmers fretted that Exchequer was a "dead horse, now" and worried that the I.X.L., too, might "collapse for want of funds."[82] When a critical letter arrived from London he erupted in anger, threatening to resign. He apologized to Syme the very next day. "I wrote you yesterday *in rixa*," Chalmers admitted. "Today the venom of your sting has lost its power."[83]

Syme finally convinced Chalmers that the best way to improve the companies' chances of success was to return to London and lend his efforts in person—who, after all, could better explain what he had seen in the mines?

Reluctantly, Chalmers agreed. On January 24, 1872, he waved goodbye to his brother and Lewie. The timing was terrible; it was the middle of winter, and a 'snow blockade' delayed his cross-country train for several weeks *en route*. Chalmers finally reached New York City exactly a month after leaving Alpine, departing for London that same night aboard the White Star Line's *Atlantic*.[84] He would not get back to Alpine until June 8, 1872.

Alone Again

In Chalmers's absence, brother John kept up a running correspondence with the London directors, meticulously reporting the total number of feet made each week in the Exchequer Tunnel. But Lewie's health failed that spring. He passed away on May 10, 1872, while his father was still away in London. Had he lived three more weeks, he would have celebrated his twenty-first birthday.

Lewie's uncle John did the needful, arranging for the young man to be buried in a quiet spot near the creek, not far from house they had been sharing at the Davidson Mill. All three shifts of miners were excused from work that day so they could attend Lewie's funeral.[85]

Lewie, unlike his father and uncle, had been well-thought-of by the miners. One former mine superintendent wrote tartly to a friend: "It would not be such a loss if it was Captain John."[86]

Chalmers returned to the house at Silver Creek in June, a month after his son's passing. He arrived fortified with "lots of money" and bringing "some very rich presents," according to housekeeper Mary Hanson.[87] But his heart remained heavy. "The House looks very dull without poor Lewis," Chalmers confided to Syme. "I can't turn without being reminded of him."

Lewie had already been weak from consumption when he arrived, but it must have been especially painful for his father to know that it was work for the company that brought on his final illness. "It appears he got a severe cold going down to Bulliona in the snow when I was away, [to] receive some freight for the mine," Chalmers wrote. "Altogether my poor boy's death has upset me considerably."[88]

Perhaps as a way to deal with his grief, Chalmers threw himself into work for the mines again. Within weeks of his return, he was able to announce that he had relocated a fresh claim for the company in Scandinavian Canyon; negotiated for two more claims (the Pine Tree and Adolphus); and dispatched two tons of ore to Monitor for wet crushing at the Tarshish Mill.[89] He had arranged "very favorable purchases of good, suitable machinery for the hoisting department and a turbine wheel" while in New York, and by August 1872, those purchases were on the way.[90]

Sadly, the original headstone for "Lewie" (Lewis Chalmers Jr.) was stolen. This reproduction was made from a rubbing taken from the stone before it disappeared, and is preserved at the Alpine County Museum. (*Dustman photo*)

That August saw Chalmers busily engaged in putting up his new hoisting works and remodeling the outdated Exchequer Mill. He declared himself happily "over head and ears in work."[91] But brother John had had enough. After "spending two and a half years here fruitlessly," as Chalmers put it, John was ready to go home. It was not only that John "couldn't live on promises," though Chalmers gave him that face-saving excuse. In truth, the brothers were not getting along.

"His ways are not my ways exactly, although I have nothing to find fault with, really," Chalmers wrote to Syme. "But I don't like waiting breakfast in the morning, and I want my table immediately after [for] writing purposes, etc., etc. John is a fine, honorable fellow, but not the kind of man that can take things as they come and be content." In a subsequent letter, he added, "The facts are that he was sick of his life here, and I wished to be alone."[92]

Chalmers scraped up enough cash to send John home in September 1872, though he had to borrow to do it. John, unlike his older brother, had never been truly bitten by the mining bug. "I know you hate this country, and have not only no taste for, but a positive aversion to mining pursuits," Lewis wrote John a few months later.[93]

Lewis Chalmers was now alone again in the house on Silver Creek—and declared he liked it that way.

5

A Year of Disappointments and Disasters (1873)

Chalmers may have thrown himself into work as a way to cope with the grief of losing his son, but his burst of intense concentration paid off for the companies he was managing. By the closing weeks of 1872, new hoisting works were being installed at both the I.X.L. and Exchequer mines, and a pair of identical 30 × 50-foot frame buildings were going up to house the hoisting machinery.[1]

"The last shingle [at the I.X.L. building] will be on tomorrow," Chalmers announced proudly on November 8, 1872, in a letter to Syme. Its twin at the Exchequer would be finished shortly thereafter, though it would take a bit longer for its interior works to be installed. However, by mid-December, both sets of hoisting works were essentially complete.[2]

Chalmers was also busy retrofitting the old Davidson Mill. This had originally followed the Freyberg "barrel amalgamation" system, dry-crushing the rock followed by roasting in reverberatories. Chalmers had planned initially to add a White's roasting furnace, but he discarded that idea after the Schenectady mine in Monitor gave the White's system a try and pronounced it "a miserable failure." Instead, the Monitor folks turned to a wet process using bluestone and salt. This proved far more successful and, as Chalmers enthused, was the same system "now practised almost exclusively on the Comstock and similar ores."

After paying a visit to Virginia City to learn more about this "wet process" first-hand, Chalmers resolved to dispense with roasting entirely and to use the money "saved" to boost the capacity of his Exchequer mill even more than originally planned, increasing it to 16 tons capacity instead of his original goal of 12.[3]

By late December 1872, Chalmers had blown through $7,594 of Exchequer's funds to convert the old mill to wet crushing. The original mill buildings had been repaired and reconfigured, two new mortars and blocks added to replace broken ones, and the old stamp battery rebuilt and rebraced. Four Hepburn steam-chested pans and two 8-foot iron settlers had been added to the configuration, plus a new cam shaft, clean-up pan, and other parts.[4] The retrofitting of the mill was still not quite complete as the New Year rang in, but "[b]y the first of April I hope to be able to congratulate you on the possession of a perfect sixteen-ton Mill, equal to any of the same capacity in the State," Chalmers wrote home encouragingly on January 1, 1873.[5]

Hoisting works at the Exchequer Mine in Scandinavian Canyon. (*Huntington Library, Withington stereoview, ca. 1876*)

Hoisting works at the I.X.L. Mine in Scandinavian Canyon. (*Dustman collection*)

After discovering that oxen could only accomplish one round-trip from mine to mill per day, he traded in his ox teams for four stout horses, capable of hauling twice as many loads. He also snapped up two used quartz wagons, dubbing them "almost as good as new, at one-fourth of the new price." With that new ore-hauling system in place, he declared himself ready to haul 16 tons of ore per day to the refurbished mill.[6]

If things seemed to be progressing well at the mines in Alpine, however, back in London Chalmers's family was in dire straits again. Sister-in-law Eliza had become so desperate for funds she had presented herself at the company's office in London, seeking aid. "I have told her frequently never to write or go to the Office," an exasperated Chalmers confided to Syme, "but she is as wrongheaded as a mule."[7]

Chalmers himself was feeling equally pinched. "Can you raise a £100 or so for me," he begged Syme in November 1872, describing himself as "hard up," and complaining: "I receive the pay of a powder money and do the work of an experienced gunner and a full gun detachment." His patience fraying, Chalmers added an ultimatum: "[W]hen money fails, I stop work; [I will] write one letter for funds and wait for a reply. If no money or guarantee from your side, I leave."[8] Although he received no immediate relief for his own tight purse, Chalmers did wangle a promise from the company to furnish Eliza with a much-needed $500. "For goodness sake, take care of it," he admonished her.[9]

As 1873 dawned, Chalmers continued to do what he could to advance the investors' interests, despite his lack of funds. He negotiated a purchase of the Adolphus and other nearby mining claims in Scandinavian Canyon, and put his legal skills to work defending Exchequer's title to a disputed portion of the Lady Franklin lode.[10] He also touted the Leviathan copper mine in Monitor, urging Syme to "take this up as a private adventure":

> Go into this thing at once *con amore*.... Don't lose this chance of a sure £2,000 a year ... Within four months of commencing operations, Mr. R[ickard] assures me he will be making a profit of $3,000 a day, and there is any amount of ore in the mine. He says there is on the dump enough ore to produce $20,000 worth of bluestone, the profit on which would be $10,000.[11]

Chalmers tried his hand at photography again, though once again, his efforts fell short. He had attempted a dozen photos to send home, he lamented, but they had all been overexposed and were "of no use."[12] As always, Chalmers kept up a vigorous correspondence, though he declared he had little time for letters "between mines and mill, legal proceedings, drawing plans, searching records, obliged to go to San Francisco and Sacramento, etc."[13]

As January 1873 gave way to February, and February gave way to March, the spiffed-up Exchequer Mill remained "not yet finished." Not that it really mattered all that much; there was no way to supply the mill with ore anyway, with 15 feet of snow clogging the road in places.[14]

Inside the Buckeye tunnel, the miners continued to work, but water had begun penetrating the shaft, making working conditions miserable. Chalmers was forced to cut his shifts from ten to just eight hours a day and to hire additional miners. "It is so awfully wet there, the men could not stand it, in water (drip[ping] from above, cold ice water) for ten hours, as formerly," he explained, justifying the added expense.[15]

His brusque management style was also causing difficulties. Chalmers found himself in "bad odor" with his carpenters after chastising them for smoking on the job. "Caught one at it the other day," he reported with evident satisfaction, "and gave him fits before the rest, which had good effect." Unhappy as this must have made the men, Chalmers seemed to revel in his reputation as an unflinching taskmaster. "I am afraid I am thought unreasonable by both foremen," he confessed with a touch of pride, "[as] I am not quite satisfied with the work done. I may possibly want too much."[16]

He refused to tolerate slackers, promptly dismissing a foreman caught whiling away a snowy day. "I had occasion to go to Silver Mountain on snow shoes [skis] ... in a snow drifting day when he didn't expect me, and found him there in the middle of the day, playing poker with a Chinaman and drinking," sighed Chalmers to his brother. "I had no alternative but get rid of him."[17]

Chalmers's sharp tongue likely prompted many of his staffing headaches. In one unintentionally humorous letter to a housekeeper poised to depart after Chalmers had apparently offended her, he hinted at remorse but offered no actual apology:

> Mr. Chalmers is sorry Mrs. Cummings should think of leaving so hurriedly. Mr. C. will be obliged by Mrs. Cummings remaining and doing her best until he can get another Housekeeper to supply her place.
>
> Mr. C. did not wish Mrs. C. to leave, but he did wish she would try and make the men comfortable.
>
> Mr. C. will get another housekeeper as soon as he possibly can.
>
> Mr. C. cannot settle with Mrs. Cummings 'till the first of next month, so she had better stay 'till then.[18]

On March 21, 1873, Chalmers penned a lengthy (and glowing) letter to Prof. R. W. Raymond, the U.S. mining commissioner in New York, summarizing both Exchequer's mining efforts and similar activity at nearby mines. Despite all the difficulties he had encountered, his accomplishments were indeed brag-worthy.

Since commencing work at the Buckeye and Accacia Mines in 1870, Chalmers reported, his team had "driven their Main Tunnel 900 feet, sunk over the Ledge 140 feet from [the] Tunnel, raised 128 feet of Air Shaft, and run 912 feet of drifts." A 24-foot hoisting frame with double sheaves had been erected, driven by a Bacon's two-cylinder engine, allowing the hoisting to be "done to perfection, smoothly and evenly, at the rate of 100 feet in thirty-four seconds." A Blake's steam pump was in place to rid the shaft of water, removing some 8,700 gallons every twenty-four hours, most of it surface water "from thawing snow." A tramway had also been erected to ferry ore from the works to a dump on the road, some 240 feet below.

In addition to Exchequer, a separate English company had been formed to acquire the I.X.L. Mine, some 1,700 feet lower in Scandinavian Canyon. The I.X.L. had been the discovery that "gave birth to the Town of Silver Mountain," he observed, producing an estimated $50,000 in years gone by "from a very limited Stope." Ore from the tunnel still featured tantalizing pockets of fine ruby silver, he quipped, adding: "An expert cannot distinguish some of this ore from Yellow Jacket ore on the Comstock Lode." The company had its own mill site beside Silver Creek and predicted they "will erect a mill this fall, should the shaft developments justify it."[19]

Included in Chalmers's comprehensive report was a brief mention of his now-defunct Imperial mine near Monitor. Putting the best face possible on its prospects, Chalmers observed that the Imperial's abandoned 1,406-foot tunnel was "now within 300 feet of their Second Ledge." Indications there remained "very promising," he stressed, "but the Company has suspended work until Exchequer and I.X.L. bullion shipments encourage them to proceed."[20] His lengthy report ended on a predictably upbeat note:

> Up to a recent period the mines of Alpine County have been mostly owned by poor miners and others who, unable or unwilling to work their silver, have steadily refused to sell unless at highly exaggerated prices. Tired of acting [as] dog in the manger, they have at last offered inducements to Eastern and English capitalists who found the mines totally undeveloped.... Now, I think the prospects of this Section are exceedingly encouraging and ere many months are over, bullion and bluestone will flow from Alpine.[21]

The truth, of course, was far more complicated. Beneath the hopeful words and good intentions, the real-life challenges at the mines continued. Dissatisfied with his crew's progress in the Exchequer shaft, despite "scolding, coaxing, and using every endeavor likely to accelerate the work," Chalmers finally paid off his current hands in March and started for Grass Valley to procure replacements. He returned two weeks later with a fresh foreman and four miners in tow, "all well recommended."[22]

Renovations to the mill continued, albeit slowly. "The Pan floor is half-finished; the old dust floor removed, and the engineer busy fitting pan and Settler gearing," he chafed. "[B]ut there is still a good deal to do."[23]

Meanwhile, at the mine itself, the water intrusion was growing worse. "On the 17th, 18th and 19th, we were hoisting water day and night and could only just keep it down," he reported in frustration. "The Shaft is too shallow yet to put in pumps. We should have to shift the pump and pipes too often. We have therefore to use buckets."[24]

What Chalmers did not mention in his letters was a cataclysmic event that had just taken back in Washington, D.C.—a seismic shift in monetary policy that few (including Chalmers) likely recognized at the time. On February 12, 1873, the U.S. Congress passed the Coinage Act of 1873, which was quickly signed into law by President Ulysses S. Grant. This demonetized silver by halting the coinage of silver dollars, and in future years would be dubbed the "Crime of '73" for its devastating impacts. Though the effects would take several years to be fully felt at Silver Mountain, it put the nation on the road to a gold standard—and a national depression.

Yet Chalmers's concerns were more immediate; he was once again struggling to maintain the confidence of his London backers. Lord Poulett, his erstwhile supporter, had "lost faith in both the mines and me," Chalmers acknowledged to Syme, adding with a shrug: "He ought not to have gone into Mining."[25]

It was yet another reminder of the remarkable down-turn in his fortunes. "I once had a good position, and a man they say is never twice fortunate in the course of his career," he continued ruefully, adding: "*Sed dum spiro, spero*"—but while I breathe, I hope.[26] Chalmers's confidence in the rich potential of his mines remained unchanged. "The success has yet to be proved, and God grant my money may hold out 'till I prove it. That it is there, I entertain no doubt," he assured Syme.[27]

Carte de visite of Lord Poulett, taken in Paris. (*Alpine County Historical Society, Arnot photo collection, ca. 1870*)

Fearing Syme, too, might be losing faith, Chalmers wrote a few days later to bolster his enthusiasm: "You must keep up your spirits and I will do my d[amne]dest to reward those who have trusted me.... I have got two good foremen and two excellent crews of men. If snow moves away, I shall have my roads repaired and prepare for a busy summer." As added encouragement, he attached figures supporting a potential $480 profit per day (nearly $150,000 a year) once his mill began running on the hoped-for 16 tons of ore a day.[28]

By June, it seemed Chalmers was on the brink of success at last. The Exchequer Mill had been expanded to a 20-ton capacity, and outfitted with a pair of Hendy's concentrators. Bridges between the mine and mill had been strengthened to prepare them for the coming onslaught of 5-ton ore wagons; "the team, in fact, having gone through once or twice," Chalmers observed. As he added wryly, "I can't afford to break horses' legs just at present."[29]

Those upgrades had been costly. The teams and wagons alone had set the company back $3,000, and mining supplies for the previous year had added another $3,351. Chalmers's own salary, travel, and offices expenses had consumed $4,732, and the two Hendy's concentrators had cost another $635.[30]

But shipments of ore to the mill had commenced, and the foreman was confidently predicting twenty tons could be supplied every day for at least two years, just from the ore body already in sight. By June 13, Chalmers was gleefully anticipating that he would be shipping bullion in just ten more days.[31]

On June 30, Chalmers enthused to his brother, John:

> For months back, I have scarcely known whether heels or head were uppermost. I thought the workmen, particularly the engineers, would never get done, and found myself swearing egregiously. Thank God, all is finished, splendidly done, works

> admirably as [nearly as] I can yet judge. You would not know the [Exchequer] Mill again—Like the Irishman's gun it has got new stock, lock and barrel and is as near perfect as a mill can be. I want just two things: a rock breaker and large ore house for winter; but I needn't speak of this at present.[32]

But fate would prove fickle indeed. The summer that had begun so encouragingly would soon give way to despair.

A boiler failure in early July brought the mill's operation to a frustrating halt. "I was so delighted when I got started, everything ran so beautifully," Chalmers fumed in a letter to Syme.

> [T]he Battery running 100 [r.p.m.] with Engine only [turning] 78 and steam [in the boiler] at 50 [pounds], that I never anticipated any difficulty in driving my pans when loaded. They were then running loose, with mullers up, as is done when machinery in a Mill is newly started. Well, I shall say no more about it except that by selling my old Boiler at I.X.L. [mine] I have paid so much of the new one as that I shall have it here on Wednesday, and hope to be running again by August first (if my men stay, which I doubt).[33]

It took another two weeks, but by July 22, Chalmers was able to report that he had gotten the new boiler installed at the mill.[34] That same letter also enclosed a pair of remarkable photographs documenting his improvements to the Davidson Sawmill and the adjacent Exchequer Mill, along with detailed descriptions:

View of Chalmers's compound at Silver Creek in July 1873, including sawmill, blacksmith's shop, and roasting furnace. The roof of Chalmers's house is just visible near the tall smokestack. The hill beyond the mill was denuded of its timber for both fuel and lumber. (*Alpine County Historical Society*)

The front view shows, on the extreme left, the Company's Boarding House ... close by the flume, [with] penstock and Saw Mill to the right; in front of the Saw Mill the logs on the Skids, [with] Log wagon and oxen just arrived with a load of logs. [T]he old Blacksmith's shop in the center of the picture, in front; near a ladder, the place I have excavated for the new Boiler; to the right of that, the water cistern and gable of the Engine room, on the roof of which you will see the hole through which went the old Boiler smoke stack. In rear of the ladder, the timbers framed for new Building. At right angles to the Engine room wing is the new ore house; back of the Engine Room wing the new Pan floor wing; and at right angles to the Pan floor wing, the Settler and Concentrator Wing. Right of the new ore house is the roof of the old one; [and] right of that, the Roasting Furnace and Chimney. In front of the Chimney, the Team Stable and Corral, and to the right of the tall chimney [is] the Manager's house. In rear of the roasting furnace you can just see the roof of the Assay Office, and back of all, the pine-covered mountain traversed by [a] wood road.[35]

[This] view is of the rear of the Mill, taken from the pine-covered mountain before referred to. On the extreme left front is Silver Creek, spanned by a Bridge on which stands the Logging Team, driver and skidder. Across the Bridge to the right [is] the Manager's house (gable and back view), wood house and kitchen (all the same as when I went there); right of that, the Assay office with brick Chimney and two long chimney

Oxen delivering wood to the Exchequer Mill. (*Society of California Pioneers, San Francisco, Chas. B. Turrill* collection)

> tops (built by me). The low Building to the right of the Assay Office is part of the Retort house, almost communicating with the Assay Office and Melting furnaces. On the right of the Retort house and communicating with it, is the three-storied Building, the lowermost floor of which contains the Circular Shafting and Machinery, and Concentrators; the second floor, the Settlers, Clean-up pan and Amalgam Strainers; the third [floor], the Pans, Sand Tanks and Engine room.
>
> In the rear of this tall Building is the Wing with Ventilator on top, connecting the front wing in which is the Battery, and at right angles to the ore house seen in the other View. The Mill is built on the sloping side of the Mountain, which makes the front look so much more dwarfy than the rear. To the right of the three-storied building is the Saw Mill; below, the Car Tramway from the third floor of which you can see the Slabs which furnish the Engine with firewood, and on the extreme right of that Building, the other Tramway which conveys the Sawn Lumber to the Lumber Yard. Still further to the right, but hidden by the thicket, foliage of the Pines and Cottonwoods.
>
> Back of the Manager's House is the rear view of the Roasting furnaces, with their tall brick chimney, and back of that again but separated by the Toll Road to Silver Mountain, the Cattle Corral and Stable. The Town of Silver Mountain lies to the right about two miles.[36]

On August 6, the Exchequer Tunnel struck its lode, prompting great excitement. Chalmers assayed selected specimens, with results ranging from just $100 to an astronomical $2,000 per ton. In a cablegram to London one week later, he happily confirmed that the mill was running "splendidly" and acknowledged receipt of a most-welcome infusion of £1,500.[37]

Calamity Strikes

That would be the last of his happy news. "Calamity" struck in September when one of the heavy ore wagons broke down, leaving Chalmers no way to continue to supply the mill with the necessary stream of ore. Now, instead of dreaming of silver bars, he was putting affairs in order to prepare for his own departure.[38]

> [T]he fact is, the overwhelming nature of the calamity—my present inability to supply the Mill with ore—after everything, only a few days ago, looked so promising—took all the venom out of me.... I do not think that you can carry this thing on without more money, and nothing short of a miracle now in the shape of a huge deposit of very rich ore will enable me to do so without assistance, and I do not think you can get more money so long as L[ewis] C[halmers] remains manager. I have therefore to go.... I have some furniture, [a] watch, Lewie's horse, transit equipment and books which will give me money to go out and keep breath in 'til I get something to do.[39]

To the Bank of California, Chalmers offered a slightly different explanation, saying only that he had "been obliged to shut down for want of quicksilver."[40] In a subsequent letter to friend A. F. Wheeler, Chalmers confirmed that the Mill had "shut down for want of ore to supply sixteen tons per day at present."[41]

"Dispose of me, therefore, as you will," he wrote in defeat, frankly acknowledging that the company would be better off without him as, "without changing your Manger," they would have difficulty raising additional funds. "I do not wish to desert the ship because she's sinking," he insisted, "but you will find the public—like the Sailors of old—look on me as your Jonas."[42]

Writing to Lord Poulett, Chalmers admitted his chagrin at the prospect of leaving. "I have spent five of the best years of my life on this undertaking," he sighed, "and properly backed, would spend five more." His confidence in the value of the mines remained unbroken, he emphasized, reiterating that recent samples taken "with my own hands" from a drift had assayed at over $2,000 per ton. "Five years' experience has taught me that there is nothing so likely to lose a man his reputation as mining," he added, "while at the same time there is no other business that is so likely to bring one sudden riches."[43]

For a short-term cash fix, Chalmers processed ninety tons of waste rock from the I.X.L. dump—ore that had previously been "thrown aside by the former owners as too poor to work."[44] The three-day run produced a small bar of bullion weighing 222 troy ounces, worth roughly $450. This Chalmers sent to his bank in San Francisco, where he was again overdrawn, "to keep them in humour."[45] But on September 22, his resources exhausted, Chalmers paid off all hands and shut down the mine.[46]

Some London directors acidly attributed the troubles to poor management, but Chalmers was having none of it. Let them send on someone better, he retorted in a letter to his brother:

> If they have faith in the Mine but not in me, I told them in my last [letter] what to do. The man they send [to replace me] will require to be, first and foremost:
>
> (1) Thoroughly honest; (2) A good assayer of both ore and bullion; (3) A bullion melter and furnace builder; (4) A fair surveyor and engineer; (5) A good miner and judge of ore; (6) Determined and uncompromising; (7) Somewhat of a lawyer; (8) Able to rough it; (9) A good horseman and judge of horseflesh; (10) A little of an architect; (11) Fond of living alone; and (12) If possible, he should know a little of silver. I should have added, (13) A good bookkeeper by double entry, and Lastly, a man who, when he knows he is doing the best for his constituents, and that better cannot be done with the means at his disposal, will not be offended if lectured like a Schoolboy by those who don't begin to know one iota of what he knows on the subject on which he is lectured ... N.B.: The Merchants are pressing me for their Bills.[47]

By mid-October, Chalmers's overdraft had swelled to $5,000 and he was begging the bank for forbearance. "Bank and merchants getting troublesome," he urged London.[48] By late October, he seemed to have made up his mind to leave. "Everything is in excellent order about the Mine for efficient work," he wrote the company, "and I hope you have given your new manager instructions to sink another one hundred feet."[49] Still, he could not help dangling words of encouragement. "[A]ll Silver Mountain has it we have a 'big thing there,'" he added, "[and] our own men think and say you are to have a Second Comstock, a little deeper down."[50] To his creditors, he simply lied. "I am in daily expectation of a remittance from London," he wrote one supplier reassuringly. "It will be all right."[52]

Now down to his last bale of hay and an empty corn bin, Chalmers floated a fresh idea in hopes the investors "wish to risk yet a few thousand dollars on the chance of making a lucky hit." His suggestion: set four double-shifts of men to sinking on the Accacia lode for one month, add a cross-cut tunnel, run north and south on Exchequer's ore "and see if I couldn't run the mill on the out-turn." Already there were 50 tons of good ore at the mill, he emphasized, and a 20 percent dividend in July was still possible "if you work the mine properly all winter."

Yet it was not simply a "few" thousand dollars he needed for this new push but over $22,000.[52] Syme had tried to convince Chalmers to stay on the job if another $3,000 was sent, but Chalmers had dismissed that sum as "quite inadequate."[53]

Chalmers had doubts that the Londoners would go for his new plan. "If the Board don't pay up at once, I shall leave," he confided to Eliza, "go to San Francisco and see if I can get into something, and bring the chicks out."[54] Surprisingly, his ambitious proposal met with approval. By November 11, work had resumed and the I.X.L. tunnel was again pumped dry.[55] Around mid-November, Chalmers was assuring creditors the companies were "now re-organizing," noting he had received a telegram "from which I gather I shall be in funds soon."[56] Five days later, he ordered a large quantity of pipe and fittings from Treadwell & Co. of San Francisco.[57]

The bank was still being "troublesome," his men were "threatening proceedings," the boardinghouse keepers were "grumbling," and winter supplies had yet to be purchased.[58] However, Chalmers was not about to tuck tail and return home, only to face jeers, insults, and accusations of swindling the investors. He much preferred to stay and continue the fight.

"It is not that I like this country—far from it," Chalmers assured Syme. "I hate the people, and I don't much care about the country. The sort of Yankees I meet here are, to me, the most despicable men I ever met ... [But] go back to London and starve? Not quite."[59]

6

Trying Times and Tenacity (1873–1874)

By late November 1873, Lewis Chalmers was able to share a few bits of good news with his London backers. The north drift of the Exchequer tunnel had been pushed 171 feet from the cross-cut, with "some fine stones of ore taken out last week," and a nearby stope was "turning out some good ore, and shows a marked improvement in the hanging wall side."[1]

As usual, that small dash of good news was leavened with bad: his entire crew of miners had quit and decamped for Silver Mountain City the previous week. Chalmers explained:

> I am sorry to have to state that, partly on account of the inferior quality of the food given them by the Boarding House Keeper at the mine, [but] mostly on account of "no pay," all hands except the Engineers left the Mine on Thursday morning and went to Silver Mountain, sending a deputation to me. I went to Silver Mountain and saw the men, after turning out the old Boarding House Keeper at the Mine, and [after] getting Justice of the Peace [R. H.] Ford—one of the old Pioneers of Alpine and a much respected citizen, to take the Boarding House—succeeded in getting the men to return to work on Saturday morning.[2]

Chalmers was hardly alone in failing to pay his miners on time. Some mining companies strung their workmen out even longer. Not long after the Exchequer mutiny, the Silver Glance at Monitor was six months late paying their employees, while the nearby Monitor and Northwestern mine lagged seven months behind in their payroll.[3]

It may have been a fairly common practice. "But this 'No Pay' business won't do," Chalmers chided his London backers. It was, after all, approaching Christmas, and most of the miners had "wives and families in Grass Valley," he noted, and "especially at this Season, wish to send them money."[4] Purely sentimental considerations aside, Chalmers was increasingly worried about liens, confrontations, and sabotage to the mine. "They are bound to give trouble if another month passes without paying them two months," he hinted darkly.[5]

The financial hitch this time was more than the usual difficulty of attracting fresh investors to the consortium's projects. Like the ill-fated Imperial, the Exchequer

Stereoview of the Exchequer Croppings, with two shifts of miners and a possible foreman (right). Portly figure second from right is likely Chalmers himself. (*Yale Collection of Western Americana, Beinecke Rare Book & Manuscript Library, Withington stereoview ca. 1876*)

and I.X.L. were now undergoing reorganization, a time-consuming legal process.[6] Adding to Chalmers's discomfort, he remained badly overdrawn with the Bank of British North America and the bank manager was threatening to bounce his checks. Rather than apologize for the shortfall, however, Chalmers responded with a cheeky letter, minimizing it as a petty inconvenience and flaunting his own previous high-flying position in Scotland:

> When I was Agent for Bank of Scotland, Fraserburgh, I did not get alarmed at an overdraft of £1000 on such an account as the Exchequer Company's has been to you. The Company are reconstructing, and assure me of all the money I want in a very short time; so that you need not have any fears. Of this I personally assure you, that I would not draw on you further than I have done ($5,000) until we are in funds; but I hope you will not refuse any of the Cheques I have drawn, as it would hurt us here materially.[7]

With no financial relief in sight from London, Chalmers fell back on his own resources. He managed to borrow $1,350 on his own credit—enough to cover one month's back wages.[8] This staved off a full-scale revolt at the mine, though four hands still threw in the towel and set off for Grass Valley "the moment they got it." Chalmers quickly hired on replacements, but morale remained low.[9]

> As stated in my last [letter], I paid the men one month's wages borrowed for the purpose, [but] not before one-half of them had again left the mine because they were not [fully] paid. Had I allowed them to go—which I regret I couldn't—they would have attached, and I could not have hired others. Men working in this way don't work. They know the position and take advantage of it, and nothing annoys me more than to feel that the scamps are shirking in spite of me. The Shirkers are all marked, however, and the moment I am in funds, they go.[10]

With his signature blend of confidence and offended dignity, Chalmers sent a pointed letter to Syme on December 3, again threatening to find other employment if funds were not forthcoming.

> Let me [get] down 100 feet and run the Accacia, and my Exchequer Mill will run on Exchequer Ore. I know it. Next week if this snow storm don't baulk [*sic.*] me in some way, and I can get men (which as yet I have failed in doing, owing to our impecunious condition) I will commence a run which I hope will result in driving some of this stuff—this infidelity—out of your sceptical [*sic.*] heads, and opening the strings of your well-guarded purses ... I am here suffering daily annoyance and discomfort because I believe in the Mine; but Syme, unless money is immediately forthcoming to pay the men, I will not stay. I cannot pay my [household] Servant, and by and bye won't be able to get provisions.[11]

The situation was fast becoming precarious. His supplies of hay and oats were almost exhausted, and there was no likelihood of getting more. The new boarding house keeper was demanding his salary and, "if not paid, will throw it up, which means shutting down the mines," Chalmers warned London. "Every day I have duns, and can tell them nothing. You know the result of this."[12]

Unfortunately for Chalmers, the early snowstorm he had mentioned in his letter of December 3 did not abate. Four days later, the snow at Chalmers's house at Exchequer Mill was 3.5 feet deep, with 5 feet "on the level" at the mines up in Scandinavian Canyon, and the road to the mines buried under 6-foot drifts.[13] That brought ore deliveries to the mill to a screeching halt.

> On Monday I sent up the Team [of] ten horses empty, merely to break the road, but the snow was so deep I got down only a few sacks of ore, and since then it has stormed so I couldn't send the team.[14]

Compounding Chalmers's worries, he had not yet been able to purchase winter supplies. It was not for lack of trying; he had attempted to place a large order with the Sacramento firm of Adams & McNeill on December 1, spanning everything from staples like rice and flour to luxury items like canned strawberries and sewing machine needles. Yet with no money in hand, Chalmers had sought three months in which to pay.[15]

He had also attempted to place a large mining supply order with Treadwell, a San Francisco supplier, listing candles, picks, blasting powder, car wheels, nails, pipe, and other items for the mines. In that letter, too, he casually noted that credit might be required. "I may pay [for] these in thirty days, but I want them at ninety days in case I may not find it convenient," he wrote smoothly, promising to add the "usual interest" if more time should prove necessary.[16] Seemingly indifferent to the impact on his reader, Chalmers had blithely included a request that Treadwell include a good amateur's foot lathe in his order, "for my own amusement."[17]

Businessmen that they were, the Adams McNeill and Treadwell managers both refused to fill these large orders on Chalmers's easy terms.[18] Now desperate for supplies, Chalmers waxed creative. He dispatched a new, larger order to Adams McNeil, adding the desired mining supplies to his original food order. Along with this massive list he enclosed a token $200 toward the company's past-due account, and proposed text for a telegram he suggested Adams wire to London, demanding payment. This, he assured the company, "will bring you $1,000 to pay what I owe for the [Exchequer] Company and part of the accompanying order," with the balance promised in sixty days. "Hurry on the freight in any way by rail," Chalmers urged, adding: "I do hope you will ship via receipt, as if not, I shall not be able to get my winter supplies to mines at all, and so have to shut down."[19] That was a thinly-veiled threat, of course. If the mine shut down, Adams McNeil could kiss the rest of their outstanding balance goodbye.

Chalmers shot off a companion letter to his London associates to prepare them for the coming demand from Adams McNeil. And once again, he ended with a threat: "If you have sent this [$1,000] I will get the supplies; if not, I must shut down.... The supplies in both mines are fairly exhausted."[20] To Chalmers's great relief, a telegram arrived December 14, promising to cover Adams McNeil's supply bill. Better yet, it advised him that a welcome infusion of £500 was on the way and hinted at the possibility of "more next week."[21]

The money did arrive as promised a few days later, though it still was not nearly enough.[22] "Of course this will be a help, but what about I.X.L. with three months' [pay] due on the last day of this month?" Chalmers fussed. "I cannot get work out of men when they know I can't pay them."[23] When a second infusion of cash failed to arrive by Christmas Day, Chalmers complained to Wheeler: "I am running on debt every day."[24]

To Chalmers's chagrin, the San Francisco bank credited his account with just $1,440 for that recent £500 deposit. He sent a complaining letter to the bank manager about the poor rate of exchange, written Christmas Day.[25] But Adams McNeil had at last agreed to ship his much-needed winter supplies "on promise of a Speedy Settlement."[26]

Despite the impossibility of hauling ore to the mill, Chalmers had not spent the winter months twiddling his thumbs. He continued to promote the company's mines, putting his prodigious powers of persuasion to work even while the "work" itself lay idle. In late October 1873, Chalmers had written again to Rossiter W. Raymond, the highly-respected U.S. commissioner of mines and editor of the prestigious *Engineering & Mining Journal*, asking Raymond's opinion of selected ore samples he had shipped for his evaluation. Though Raymond was in New York at the time, Chalmers added hopefully:

> Could you possibly come here for a few days and examine? You would see a new Mining District (or rather an old one being revived) ... and as it is only one day's journey from Carson, would not occupy more than a fortnight of your valuable time. It might also lead to business relationships with my Constituency in London—a very powerful one.[27]

Drawing on his legal training, Chalmers shot off a second letter to Raymond not long thereafter, tactfully suggesting a "short Rider to the Mining Act of 1872" to allow foreign companies (like Exchequer) that had purchased their mine from a U.S. citizen to patent the claim. After all, as Chalmers accurately pointed out, "the law is so easily eluded by sham conveyances to U.S. Citizens, who procure a patent and [then] reconvey."[28] It was an argument Chalmers would renew the following spring,

Portrait of Rossiter W. Raymond.
(*Wikimedia Commons, public domain*)

in a similar (and equally unsuccessful) letter to Sir Edward Thornton of the British Foreign Office, complaining about the undue hardship on British investors.[29]

Prof. Raymond evidently declined Chalmers's kind offer of a visit to the Alpine mines, but he did send a return letter praising the quality of the ore samples Chalmers sent him. It was a notable stamp of approval, coming as it did from the respected mining expert. Chalmers gleefully forwarded Raymond's remarks to the board back in London, hoping it would assist efforts to resuscitate the companies' fortunes.[30]

At the same time, however, Chalmers had also been quietly hedging his bets. He answered an ad in *The Mining Press* for a "mine and mill superintendent," touting his accomplishments over the past six years. He had successfully overseen excavation of the Imperial's 1,406-foot-long tunnel, he pointed out (without bothering to mention its absence of silver); erected a "very fine mill" and two sets of hoisting works; and more recently, pushed 3,000 feet of tunnel and 340 feet of shaft, "besides a deal of other work." Not shy about his expectations, Chalmers listed his requirements were he to take the job: a salary of $5,000 per year with at least one year's employment guaranteed; a free house; and traveling expenses. He would expect a raise after twelve months, he noted, "[s]hould I give satisfaction."[31] Perhaps unsurprisingly, he did not get the job.

As 1873 dragged to a close, his miners were still pursuing the plan Chalmers had suggested: driving the Exchequer tunnel on the lode, with periodic drifts North and South.[32] But nothing was turning out the way he had hoped.

The company's debt had only continued to escalate. Including December's unpaid payroll, liabilities now totaled a whopping $18,964.54, according to his year-end summary—more than $440,000 in today's dollars.[33] Although 150 tons of good ore had been extracted from Exchequer, the rock from the once-fabled

I.X.L. had proven disappointing. Worst of all, the amalgamator Chalmers was expecting to run the now-refurbished mill was confined to bed with erysipelas (a painful bacterial infection), and Chalmers had not been able to find another amalgamator willing to travel.[34] One can almost hear the frustration in Chalmers's words to his brother, just before Christmas: "It is almost impossible to get a good man to come up to this snowy region for a short job, and when you do, you have to pay high wages and traveling expenses both ways."[35]

Meanwhile, weather continued to delay both ore and provisions. "[S]ince Monday, it has snowed incessantly, making the five feet of snow we had at the mine [now] eight [feet], and effectually stopping hauling," he fretted on December 17.[36] The team attempting to deliver his much-needed supplies got stuck between Carson City and Genoa, and that proved to be only the first of its hurdles.[37]

> My teamster had to leave my wagon at Markleeville and borrow a sleigh. Stuck at Bulliona, came on with his horses Thursday night, went back Friday and brought his Load [up] at two trips; went up to go to the Mine on Saturday, stuck half-way. Returned on Sunday and made two trips to the Mine. To-day he stuck at the Pittsburg Mill this side of Silver Mountain until relieved, but I doubt if he'll make the mine tonight. It blew a perfect hurricane last night.[38]

As for mining supplies like candles and powder, Chalmers was still waiting. On Christmas Day, he pleaded with Adams McNeil to hurry them along:

> I am sorry you have not sent the supplies [powder, candles etc.], as the Road which cost me $150 to clear of snow may be blown full again before I get them up. Were I not sure of ultimate payment, I would rather cut off my right hand than be any party to order the goods from you or anyone else. My friends hold a very high position in London and know what I am doing.[39]

The snow removal he mentioned had been performed by Chalmers's own mining hands, a task they detested. When two of his miners flatly refused to shovel snow, he simply fired them. "Discipline must be maintained," he shrugged.[40] It likely did not help discipline that "the whole Comstock range [was] now exciting much interest and high expectations," with "quite a lively time in stocks," as a mining contemporary noted.[41] Word traveled fast in mining circles, and Chalmers's miners had likely heard the news and may have been dreaming of greener pastures.

To Chalmers's disgust, Syme, normally a staunch supporter, had begun admonishing him to be more grateful for everything his London backers had done to date. That criticism clearly struck a nerve. Chalmers retorted:

> What have these people done for me? If I must tell you, I will. They made me manager of this Company at a salary less than is given to any manager on the Pacific Coast—now not paid. They get for that salary work that is not done by any other manager on the Pacific Coast—mine and mill superintendent, assayer, bullion melter, surveyor, bookkeeper, lawyer *et hoc genus omne*, and they abuse him for doing it *con amore* ... Then what have I done for them: worked for them as no other manager would, identified myself with the very life of the thing, labored for them more than if I had

> owned it myself, devoted to them the whole energy of an energetic nature and any ability I may possess because, up to a certain point, I felt they were my friends, and it pleased me, apart from the question of duty altogether, to do them a service, to repay them for trusting me.[42]

At long last, another £500 infusion arrived on December 30 in the shape of bank post bills from the Bank of England. It must have lifted Chalmers's spirits slightly, though it took time for the transaction to clear.[43]

In early 1874 a series of glowing mining reports began to hit the pages of the local *Alpine Chronicle*—likely penned by Chalmers himself. A "rich strike" was reported in mid-February at the Exchequer, with the ledge gleefully described as "fully six feet in width" and expected to "pay handsomely."[44] "Good progress" was noted the following month, accompanied by more glorious hearsay: another rich body of ore was now visible in a drift.[45]

But as usual, a handful of setbacks accompanied such positive news. The Exchequer's boarding house caught fire in late February 1874, evidently from an exploding lamp oil. Seeing the kitchen ablaze, the Chinese cook roused the sleeping miners with screams of "fire!" Men leaped from the boarding house windows in panic—some finding a relatively soft landing in deep snow, others tumbling more painfully to the rocky cobble of the dump.

Luckily, damage to the boarding house proved minimal in the end, with only a bit of charring to the kitchen's woodwork and a few windows broken.[46] But the prospect of disaster may have roused at least one unpaid creditor to action. Supplier Adams McNeil—fully paid by Exchequer but still owed for supplies by I.X.L.—filed an attachment. The sheriff seized I.X.L.'s moveables to secure the creditor's claim, locking candles, powder, and other mining goods in a building at the nearby Lady Franklin Mine. That turned out to be a very bad idea.

Yet another fire, this one in May, consumed the sequestered I.X.L. stash, burning up twenty boxes of candles—a $125 loss.[47] A keg of powder consumed by the inferno made things "hot and lively," as the *Chronicle* put it. Luckily, the flames did not reach an additional fifty kegs of blasting powder sitting just 15 feet away. So much for Adams McNeil's attachment.[48] But so much, too, for those much-needed mining supplies.

As if that were not disaster enough, another of Chalmers's boilers gave up the ghost that spring, "its lungs having given out," as the newspaper said.[49] Foreman James Champion suffered a serious accident on his way to the I.X.L., losing his footing on the road and tumbling some 300 feet down the mountainside, though luckily, he survived the ordeal.[50] In May, rushing water from spring snowmelt carried off the dam at the Exchequer Mill.[51]

By the end of May 1874, ore was again making its way to the mill.[52] Recognizing the value of good publicity, Chalmers shared promising specimens of ruby ore with the local newspaper.[53] He was rewarded with encouraging coverage. "We are satisfied the Exchequer has a great future before it," the editors enthused in mid-June, adding the following week: "The Exchequer quartz wagon passed our office to-day, loaded with very rich ore."[54]

Even the I.X.L.'s prospects finally seemed to be looking up. In a report submitted for the 1874 edition of Raymond's *Mineral Resources* compendium, Chalmers predicted:

> Two hundred feet more of driving, at a cost of from $12 to $14 per foot, will bring us under the perpendicular of this ore-body, and should it dip north, as I think it does, a few feet more, into a mass of ore which, even at this shallow depth, will redeem some of my prognostications as to the great value of this lode when thoroughly opened up.[55]

That September, Chalmers wangled a glowing report on the Exchequer from the noted R. W. Raymond himself. Dated September 16, 1874, the report offers a bird's-eye view of the mine's various improvements as of that date, though Raymond himself had never actually visited. The main tunnel now stretched 811 feet long, with a separate tunnel driven on the Accacia intersecting a second lode. Hoisting works had been erected at the Buckeye claim, and near the manager's house stood a sawmill, an eight-stamp mill with four Hepburn pans, a stable and blacksmith's shop, and other buildings.

Raymond estimated the cost of future mine exploration at $48,880 and recommended that $34,336 "ought to be expended on the strength of present indications" to add eight more stamps and a new Stetefeldt furnace to the mill. Best of all from Chalmers's perspective, Raymond expressed confidence in Chalmers's "good judgment," business ability, and devotion to the company.[56] All that was needed now to close out the year on a high note was the production of a few happy silver bars, but the ore itself was not cooperating.

That September, Chalmers was forced to report that "our ore, while increasing in grade, had become quite unmanageable without roasting."[57] With no roasting furnace yet built at the Exchequer Mill, he turned to a neighboring mill for assistance—only to fall victim to what Chalmers swore was a swindle.

The Monitor & Northwestern Company's mill in Monitor was equipped with a McGlew furnace, and to prove the value of his ore, Chalmers shipped 50 tons of Exchequer ore there for reduction. Although the ore was only moderately rich at $72.96 per ton, based on Chalmers's own assay this quantity should have yielded $3,648.80 in silver and gold. Subtracting $750 in working costs at the mill ($15 per ton), he expected to be left with a profit of more than $2,500.

"What was my surprise, therefore, when I did extort a return from them, to be informed ... that the gross produce was $653.54," he spluttered. That meant he owed more in mill costs than the ore had produced.[58]

Chalmers was convinced that foul play was afoot. The mill foreman, he believed, had deliberately withheld salt during production to "make the run a bad one, knowing how much depended upon it, and believing that the [Exchequer] Shareholders, disgusted, would abandon the mine and the lienors could then be got to dispose of their interest for a small sum."[59]

He certainly had grounds for suspicion. The same day the Exchequer's ore had been worked, "two large bars of bullion .980 fine were exhibited in Monitor," although these were said to have come from a different mine. But the day after that, Chalmers added, the mill foreman "started surreptitiously in the Superintendent's buggy, took the bars with him, and had them coined (also surreptitiously) in Carson Mint."[60]

Whether Chalmers actually had been cheated we'll never know for sure. The Silver Glance mine, the ostensible source of those two silver bars, was rich enough to have produced at least one previous silver brick.[61] But though the timing was certainly suspicious, Chalmers was lawyer enough to realize he didn't have sufficient evidence to prove chicanery.

Now left with ore he couldn't successfully reduce and no bullion from his recent run, Chalmers found his prospects again looking grim. "All the men have left except two, whom I have been able to persuade to stay to keep out the water," he wrote, noting that he expected "instructions" at any time—likely an order to stop.

Even so, Chalmers put the best possible spin he could on this latest disaster. Certain that the investors would accuse him of having been overly optimistic, he puffed: "I admit my disposition is not melancholic, and that had my temperament been otherwise than sanguine I could not have put up with all I have had to endure in sticking to this thing and working it without money so long." Yet even this most recent disappointment had a bright side, he insisted. "I now maintain that I have *proved* to you that your ore will pay."[62]

7

Exchequer Gives Way (1875–1877)

Sadly for Chalmers fans, the letterbook for his personal correspondence from 1875 through early 1876 has gone missing. There's an even wider gap in his business letters, with the next known volume of "Officials" not picking up until 1878. But luckily we have been able to fill in some details of this period from other sources.

Newspaper reports confirm that two visitors from England descended on Chalmers in the spring of 1875 for a visit. These were likely emissaries from the London consortium, eager for a first-hand glimpse of its mining claims and a 'second-guess' look at Chalmers's management. Londoner Algernon Joy spent over a month with Chalmers, reporting back approvingly on Chalmers's "faith and devotion to the mine." Chalmers's frugal stewardship was confirmed with an added dose of pity, describing the formerly well-to-do manager as "cut off from civilization [and] the society of his equals," getting by with only the "bare necessaries of life."[1]

Investor W. G. Romaine of Old Windsor, England, too, came out to inspect the Exchequer in 1875, sailing for New York in late June on the S.S. *Abyssinia*. His itinerary called for a stop in San Francisco to chat with banker William C. Ralston before forging on to visit Chalmers.[2] Romaine appears to have been a savvy and well-connected investor; by 1880, he owned shares in the Comstock's celebrated Consolidated Virginia. Although Romaine's verdict on the Exchequer has been lost, he, too, was apparently pleased; he would continue to correspond with Chalmers in later years.[3]

On the national front, 1875 brought more upheaval for silver interests. The demonetization of silver in 1873 had been bad enough. But in a catastrophic flurry of good intentions, Congress attempted to further "strengthen" U.S. currency with the Specie Resumption Act, passed January 14, 1875. This granted greenback holders the right to redeem their paper money in gold, a move intended to prop up trade. Unfortunately, it proved yet another blow for silver interests, reinforcing the dominance of gold.[4] Where once the two metals had traded at a predictable ratio of fifteen or sixteen to one, now the relative value of gold began to rise. The shift was slow but inexorable; twenty years later, it would take 30 ounces of silver to equal a single ounce of gold.[5]

Workmen at the Exchequer Mills in 1876. The building behind was likely the sawmill, adjacent to the quartz mill. (*Huntington Library, Withington stereoview*)

Alpine's silver mines, never terribly prosperous to begin with, were hard-hit by this aggressive embrace of the gold standard. Predictably, interest in the local silver mines started to wane. Before long, Silver Mountain City's population was ebbing as well.

Silver Mountain City had proudly held the title of official county seat since Alpine County was created in 1864. Other towns had occasionally sought to wrest the honor away over the next decade, but Silver Mountainites had always prevailed. In September 1875, however, a fresh petition was circulated to put the county seat issue to a vote once again. Silver Mountain was no longer a thriving mining camp, opponents contended, and its winters were far too harsh. Perched high in the mountains, the town was apt to be cut off for months from the outside world when winter snows arrived. It was hardly an ideal spot for a seat of government.

When balloting took place on October 20, 1875, the contest was not even close. Out of 172 votes cast, Markleeville received a resounding thumbs-up with 129 supporters. Silver Mountain City lagged badly behind, with just thirty-five stalwarts in its favor.[6] On November 1, 1875, the change became official: Silver Mountain City was out as the official Alpine County seat; Markleeville was in.

Living as he did less than 2 miles from Silver Mountain City, Chalmers must have been aware of both the town's waning population and its embarrassing loss of county-seathood. But if he gave it any thought, he likely brushed such shifts off as inconsequential. Resuscitation of local mining when his projects came to fruition would surely bring a return flood of residents streaming back to Silver Mountain. Besides, Chalmers was busy enough with his own projects that he likely did not worry much about matters of county government.

Antoinette Arrives

For Chalmers, the most significant event of 1875 had nothing to do with monetary policy or county seats or even mining. Instead, it involved a pretty, young housekeeper who arrived to brighten his house at the Exchequer Mill that fall.[7]

Born in Shaw's Flat and most recently a resident of San Francisco, Antoinette Laughton had lost her husband only a few months earlier. She was about twenty-two years old when she came to work at Silver Creek as Chalmers's housekeeper, bringing her one-year-old son, Henry. Chalmers, now fifty, was more than twice Antoinette's age, but her youthful energy proved a welcome change for the long-time bachelor.

So how, exactly, did Antoinette get the job? Local scandalmongers hinted she had once worked in a dance hall at Virginia City. Others suggest Chalmers's previous housekeeper told Antoinette about the position.

However their paths initially crossed, by May 1876, Chalmers was doing his best to keep his new housekeeper in good humor, ordering muslin and pearl buttons for her from Carson City.[8] He placed a seed order that spring for peas and sugar peas, as Antoinette apparently wanted a garden.[9]

On the mining front, Chalmers had turned his attentions toward resurrecting the moribund I.X.L. mine, which had been forfeited to the state for unpaid taxes. With help from "a few friends," he began buying up miners' liens against the mine, hoping to reorganize the I.X.L. as a "new company," and reached out to unpaid creditors like Adams McNeil, asking what they would take "in coin" to settle their debts.[10] Chalmers was also eyeing fresh improvements to the Exchequer Mill, including the addition of an O'Hara furnace, able to roast thirty tons of ore per day. "There is no furnace that requires so little care or knowledge as the O'Hara," he assured Syme.[11] If that were not enough, he was also trying to launch an entirely new mining company, the Isabelle.[12]

As spring moved on toward summer, his plans remained "at a standstill" as he awaited developments—and money—from London. "My dear Syme, for God's sake write me and tell me how you are," he urged in late May 1876. "Say also if money is forthcoming to meet obligations and to put up furnaces, new buildings, and new flume."[13] By early June, the hoped-for funds still had not arrived. "Without money sent, I shall be shut down, so please yourself," he wrote, underlining the sentence to telegraph his frustration.[14]

Finally, money did arrive—a lot of money. The *Alpine Chronicle* described a "flood of foreign capital" into local mines, much of it likely through Chalmers's hands.[15] The long-suffering Scot must have been delighted. He finally had the resources to hire himself an assistant at $100 per month to handle such menial tasks as payroll. The man he selected, J. Thomson, came with extensive experience: "a bookkeeper [who] has had charge of the railroad works [in the past] and is now also in charge of a squad of men on the new flume," Chalmers explained in letters home. As for Thomson's background, he was "a Scotchman [who] studied law in Eden, but left at an early age for this country. Been in Mexico and is about 55, thoroughly independent and trustworthy," he reported.

Chalmers had originally hoped to give this new assistant the mundane work of keeping the companies' books, but there was an embarrassing problem: "[W]e are so far behind that I thought it imprudent for ... him to be allowed to see them," Chalmers confessed, "so keep them myself."[16]

Now flush, Chalmers seemed determined to treat himself to a bit of long-delayed luxury, beginning with his own dinner table. Perhaps he was also trying hard to impress his new housekeeper. That June, he ordered two cases "best oysters," four cases of "best sugar-cured ham," a bottle of pickled walnuts, and a 10-gallon cask of "pure, light dinner claret" from a Carson City merchant.[17] A few months later, he requisitioned two more cases of "best oysters," along with "choice" Japan tea, Indian chutney, and a box of Havana cigars.[18] He may have been stuck in rural Alpine, but at least he would eat well.

Chalmers settled into an easy domesticity with his new housemate and quickly grew fond of her. When Syme asked the origins of the little boy now sharing his house, Chalmers replied with a virtual wink: "I assume [him] to be the son of the late Mr. Laughton—husband of my present housekeeper, Mrs. Laughton, who came here a year ago. Perhaps the next one may be mine."[19]

When Antoinette left to seek medical care for her son that July and did not immediately return, he wrote in dismay:

> When can I expect you? Or are you coming back at all. The house is dull without you. Do you want more money. I have dined on just a chip and breakfasted on a cod every day since you left.[20]

By late August, Antoinette had returned but promptly got into a tiff with the female cook at the nearby Exchequer boarding house. A hilarious letter describing this domestic drama, ostensibly from Antoinette but obviously written by Chalmers himself, was sent to the woman who had arranged the cook's employment:

> You will be sorry to hear that Mary Commerford has left the Mill, going by the Big Tree Road but can scarcely reach San Francisco sooner than Friday, so you will probably have my story before hers.
>
> Mr. Chalmers said when he spoke to her in your house, she undertook to cook without help for fifteen or twenty men. When she arrived here he had nearly three times that number, and allowed her assistance—at one time two Chinamen, at another, a woman, and promised, if she stayed, to give her two Chinamen regularly. But go she would, after working eight days, which cost Mr. C. $30 and he will give her no more.
>
> Sorry for Mary, because she said she had only $15 to take her home, I let her have $40 on loan for six months, taking her note payable at that time, as she said she owed you money which she had to pay first and I was willing to wait, if she didn't find a place. No sooner did she get the money than she maligned me, if my information is correct, in the most shameful manner, so much that I did not bid her good-bye. Among other things she said Mrs. Metcalf's talk [about me] had come true.
>
> But what was my surprise to discover that, by presenting things to the men, she had managed to get $32 from them to take her home. So, as I can ill afford to lay out $40, especially to give to one who has done what she could to injure me, I have stopped payment of the Cheque and sent the note to the bank to be exchanged for it.
>
> I am sorry Mary should have behaved as she did, as Mr. Chalmers blames me for recommending her. I tried to assist her all in my power, bid her to stay a day with

me until she was rested, went up to assist her [with] cleaning, until Mr. C. came and ordered me home, as he said I was not able to scrub and work so hard. But that she was your friend, I would [rather] be too safe than too sorry.

I am now satisfied that Mary is not a good-hearted woman, else she could not have behaved here as she has done. Fancy her concocting such a story to the men as to induce them to subscribe $32 for her, while she had my $40 in her possession.

The House, [which is] to be taken down, was certainly not a nice house—but Mr. C. had her room cleaned, painted and papered, gave her his own bedding and a good bed, and seemed anxious to do what he could for her.

She is quite adept at swearing and foul language, and I am very sorry I recommended her to Mr. C.

With best regards, I remain, My dear Alice,
Your sincere friend,
Antoinette Laughton[21]

A Whirlwind of Improvements

The rest of 1876 flew by in a blur of activity. Chalmers succeeded in obtaining patents for the Lady Franklin and Silver Cloud claims, and began feverishly pushing his long-planned improvements, including adding a large O'Hara roasting furnace to the Exchequer Mill.[22]

In late August, Chalmers also signed a contract with N. D. Arnot, Sr., of the Vulcan Iron Works of San Francisco to erect a twenty-stamp mill for the I.X.L. Company.[23] It was an ambitious and hugely expensive undertaking; the contract called for $40,000 down, with another $20,000 due on completion, with the work to be finished by Christmas.[24] Translated into today's dollars, that represented an investment of more than $1.5 million.

Oddly enough, this new mill was to be erected only a short distance up the road from the existing Exchequer Mill. So why Chalmers decided the I.X.L. required its own mill—much less why the investors agreed—remains unclear. Perhaps it may have been intended to help keep the two companies' operations separate. Its twenty stamps would allow large quantities of ore to be processed quickly, and no doubt Chalmers was convinced that starting from scratch would be less trouble than the constant retrofitting and repairs he had experienced with the old Exchequer Mill. Still, it was a massive undertaking for a mine that had yet to prove it was capable of delivering a large quantity of good ore.

The new mill and O'Hara furnace were not the only improvements Chalmers was pushing. That same summer, workmen threw up a 30 × 60-foot carpenter-slash-blacksmith shop, and a new stable was planned "soon."[25] A spacious, two-story boarding house also was built to house I.X.L. workers, about half a mile below Silver Mountain City.[26, 27]

N. D. Arnot's son, Arnot, Jr., arrived at Silver Mountain about this time, likely to supervise the building of the I.X.L. Mill for his father. Arnot, Jr.'s training was not in construction but rather in law; he had graduated from Berkeley in 1869 before studying law in Albany, New York. He, like Chalmers, was a widower, having just lost his wife in 1875.

Above: Oil painting of the I.X.L. Mill and the adjacent boarding house as they looked in 1885, by Alex. J. Chalmers. Original painting at Alpine County Museum. (*Dustman photo*)

Left: Judge N. D. Arnot, Jr., *circa* 1900. (*Alpine County Historical Society, Marge Shively Collection*)

Between their similar legal backgrounds and their shared experience with losing a spouse, Arnot and Chalmers had much in common. Their personalities, however, were very different. Chalmers was a gregarious, high-flying sort, while Arnot was more staid and down-to-earth. In the coming years, business matters would sometimes join them, and just as often, divide them. They would alternate between being fast friends and determined enemies.

By late summer 1876, Chalmers had managed to restart mining in the I.X.L. tunnel after a two-year hiatus and it was not only Chalmers's mines that were doing well at the moment. The Advance Mine in Monitor was up and running with new hoisting works, after lying idle for eight years.[28] The *Chronicle* happily reported a "large influx" of miners arriving from Grass Valley—so many, in fact, that an extra stage had to be put on from Carson City to accommodate them all.[29] The "dark clouds that have so long rested on our Alpine peaks have burst," the editor enthused, with activity unseen since "the busy days of '64–65."[30]

Chalmers had his hands full. He did not even bother to aggressively contest a supplemental tax assessment for the recent improvements to the Exchequer Mine and Mill. Instead, he merely sent a terse note of complaint to the Board of Equalization, calling the proposed tax "out of all proportion to anything in the neighbourhood," but leaving the valuation issue in the board's hands, as "I have no time to fight the matter with you."[31] It was a rare concession for Chalmers, and most unlike him to back down from a fiscal fight.

In the midst of this whirlwind of activity, two more Londoners descended on him. These were guests Chalmers could hardly refuse: Frederick W. Mansell and an associate named Parrick, proprietors of the prominent *London Mining Journal*.[32] The timing was awful, but the potential for good publicity was priceless. Chalmers welcomed them with open arms.[33] The visit proved a great success. Much to Chalmers's delight, the pair later sent him photos, which he pronounced "excellent."[34] Their praise for his proposed Isabelle project also warmed Chalmers's heart: "They express themselves frequently that I am bound to be a millionaire," he later boasted to Syme.[35]

At approximately the same moment these guests were enjoying Chalmers's hospitality, workers at nearby Silver Mountain City were ripping apart the walls of the town's once-proud stone jail. A stout new log jail had just been erected in the newly-anointed county seat of Markleeville, using 7,000 feet of lumber purchased from the Exchequer sawmill.[36] Rather than abandon the original old iron jail cells, however, the frugal county fathers decided to recycle them, hauling the original heavy metal cells by wagon to their new home in Markleeville. The wagons would have passed by Chalmers's front door, though his letters never mentioned them.

August also saw three car-loads of machinery from the Vulcan Iron Works of San Francisco lumbering through Sacramento, bound (as the Alpine newspaper said) for the Exchequer Mill, but more likely intended for the new I.X.L. Mill just up the road. The equipment was so heavy it would break down several wooden bridges between Woodfords and Markleeville.[37]

Antoinette became the focus of a bit more drama that fall when her horse threw her to the ground, following that indignity up with a kick. Antoinette was "badly hurt," Chalmers reported with concern in one of his letters. The accident drew attention from not one but two nearby newspapers, one explaining that she had gotten caught in the stirrup and dislocated her right ankle. Dr. Luce of Genoa was

Photographer Eliza Withington captured this rare view of the original iron jail cells being removed from the old Silver Mountain jail during a visit to town from 26 August to 2 September 1876. She would pass away the following March. (*Courtesy of Gary Coyan*)

summoned to attend to the pretty victim, describing her injuries for the paper as "severe though not necessarily fatal wounds."[38]

Perhaps to cheer Antoinette during her recovery, Chalmers ordered two expensive cashmere wrappers for her, costing $12 (over $300 today), plus a "stylish black overshirt and jacket" for another $10 ($260 today).[39] Likely at Antoinette's prompting, he also purchased a new Del Monico kitchen range with water cistern and boiler, "complete with cooking utensils."[40]

As work on the I.X.L. Mill and Chalmers's other projects continued, word leaked out in November that the nearby Lady Franklin mine was planning to start up again soon, after lying idle since 1866.[41] Other Alpine mines, too, seemed to be undergoing a sort of small renaissance. Activity was continuing at the Advance in Monitor, and even long-abandoned claims near Fredericksburg suddenly drew renewed attention.[42] An early copper mine in Hope Valley had also begun shipping in supplies, with work reputed to be starting up "immediately."[43]

A mining engineer reporting on the Monitor-based Advance Mine that fall drew positive comparisons with the mines of Virginia City. The Advance contained a "lode of propylite so huge in its dimension as to make the Comstock appear insignificant in comparison," he wrote, hinting it could be a southerly extension of the Comstock itself:

> As this whole lode runs in a direction north ten degrees east, which is a straight course to Mount Davidson, and, as the same is traceable in the direction indicated for at least ten miles, the whole structure of this lode bears great similarity to the Comstock, there are ample reasons to suppose that there is a connection existing between them—that both even

> may be traceable to a common origin ... Whether this gives room to draw conclusions as to connections with the Comstock, I am unable to decide, but the following, a resumé of the foregoing, ought to speak for itself: (1) The great similarity of both lodes, not only in the whole lode, but in the structure of the ledges. (2) The fact that the direction of the lodes to each other coincides exactly with the course of the Alpine County lode. (3) That they both are situated in the same range of mountains, running in the directions indicated.[44]

Little wonder, then, that the Silver Mountain newspaper crowed that "[t]he stage now-a-days arrives loaded down with passengers."[45] There was even happy talk about a possible Virginia & Truckee rail extension to Alpine County. "We await with impatience the snort of the iron horse," the *Chronicle* editors enthused, "the genie that will cause Markleeville to bud and blossom as the rose."[46] It was a glorious hope, to be sure, but no such rail line ever materialized.

With multiple projects under way and the funds to push them, Chalmers was kept hopping. By mid-October, over 200 men were hard at work in his mines, mills, and wood camps, and the old Exchequer sawmill was running "night and day."[47] Best of all, positive news was flowing from inside the Exchequer tunnel. Miners reported they had struck the casing of the lode at the 400-foot level. Subsequent glowing reports described a mammoth ledge 25 feet wide, with a 30-inch vein of especially rich ore being excavated and shipped to the mill.[48]

"I am working like a horse," Chalmers sighed happily to Syme, "but am terribly bothered by low water for sawing."[49] The lack of lumber may have been annoying, but it was the least of all the troubles he had survived thus far.

Between the Exchequer's apparent success and the new I.X.L. Mill, the positive developments did not go unnoticed. As Chalmers put it in November, "There is considerable excitement here."[50] So much excitement, in fact, that adventurers were

The Exchequer Mill, undergoing renovation. Chalmers is seated on the horse; N. D. Arnot is standing, left. (*Huntington Library, Withington stereoview, ca. 1876*)

scouting out unattended claims. Jumping of mining claims and timber ranches had become "the order of the day," the newspaper reported in disgust, with false claims sometimes being filed simply for the purpose of being "bought off."[51] Chalmers's own former employee William Mercer tried to jump the Exchequer's timber ranch "for blackmail."[52] "Jumpers are on every side," Chalmers moaned to Syme a few months later. "I have had to put men in every mine to hold them."[53]

That October, Chalmers filed patent applications on the Saugatuck and Fremont, claims he had acquired for Exchequer in 1873.[54] In November, his newest enterprise—the Isabelle Gold & Silver Mining Co., Ltd.—was officially registered in England.[55] Notwithstanding everything else on his plate, Chalmers managed to procure an offer to sell the Leviathan Mine near Monitor, an opportunity he forwarded to his London friends as a possible new investment.[56] Chalmers also hadn't forgotten his earlier effort for the ill-fated Imperial at Bulliona. In December 1876, he published a patent application for eleven lodes there, collectively dubbed the "Imperial Mines & Mill site."[57]

By Christmas, Chalmers had cause to celebrate indeed. The Arnots had pushed their construction crew hard on the new I.X.L. Mill, meeting the 120-day deadline "with two days to spare."[58] The new O'Hara roasting furnace for the Exchequer Mill, too, was finally complete and ready to work Exchequer's "rebellious ores."[59] The stage headed to Carson City just after Christmas was said to be filled with carpenters "in search of pastures new."[60]

A Disastrous Mill Run

As January 1877 dawned, Chalmers was doing all he could to ensure a successful run of the Exchequer Mill, even putting on an extra team to haul ore to the mill.[61] But once again, money was running tight. "I am in debt all around here and very much bothered," he confessed to Syme in February, adding hopefully: "Exchequer Mill will run in ten days ... For goodness sake, leave me alone until I get the Mill started."[62] It would take a little longer than he expected. But the revamped mill and furnaces started up at last on March 17, with some 2,000 tons of ore on hand to be worked.[63]

Back in Ealing, his family was still pinched, and Chalmers was determined to move his daughters elsewhere to save money. He wrote to Eliza: "I shall now break up the Ealing House and send the two girls to Germany, doing at the same time the best I can for you."[64]

To meet pressing bills at the mines, Chalmers had convinced Arnot to lend him $1,050 on a sixty-day note from the Isabelle Company.[65] But expenses continued to mount. The Exchequer stable was "carried down by the rain," and a new one was needed immediately or "our horses could start to die of cold due to exposure at night," he fretted to Syme. A night watchman, too, had now become a necessity to protect the expensive equipment in the I.X.L. Mill.[66]

Chalmers deployed his usual tactics to stave off creditors, blaming the company for tardiness in remitting funds and blithely assuring merchants that funds would soon be forthcoming. To one unpaid supplier he wrote, "I am dependent on London for remittances, and lately they have been like angels' visits." To another: "I regret exceedingly that it has been out of my power to remit for your bill, in consequence of no remittance from England."[67]

The financial situation grew increasingly dire. By May 1877, the Arnots had secured judgments totaling more than $14,600, and it was not simply the new I.X.L. Mill encumbered by those liens. The Arnots had levied against all three mines (Isabelle, Exchequer, and I.X.L.); the boarding house; and even the house at Exchequer Mill in which Chalmers and Antoinette were residing.[68] Most concerning for Chalmers, though, was that he, too, could be personally liable. "The creditors, finding from the patents that the property [is] in my name, have sued me," he wrote Syme.[69]

The Arnots were not the only unpaid creditors. A $2,300 judgment was obtained by A. S. Hallidie, maker of a wire tramway, and C. B. Gregory won a separate judgment for over $2,500.[70]

Local merchants were unsure whether they should continue to honor Chalmers's supply orders. In a letter to a friend, butcher R. K. Love complained, "[I]t is the old thing, every body thinkes [*sic.*] that Chalmers will soon have money, but I do not know. He just pays me a nuf [*sic.*] to keep me hard up."[71]

The ore run at the newly revamped Exchequer Mill, on which Chalmers had gambled so much, had turned out to be "disastrous." Two of the cams broke early on, leaving each day's work "nearly two tons a day short." By late May, it was clear the I.X.L. ore he was processing was nowhere near as rich as he had expected.

Frustrated, Chalmers staunchly defend the mines' potential: "The Mines are good, but want depth," he stressed to Syme.[72] A month later, he ruminated:

> The I.X.L. has grievously disappointed me; a big, masterly-looking lode and nothing in it. But we know she was rich above, and we infer she will be so when we get down to where those riches came from. One hundred dollars per ton was paid for the rich ore in the dump ... to my own knowledge.[73]

Despite this bitter blow, Chalmers remained sanguine. To his daughter, he wrote: "As it has turned out a failure, there is nothing for me but to plod along here until I make a success of it, which can only be done by going down another six hundred feet, and which I hope the Company will enable me to do."[74]

Yet by mid-June, without the expected infusion of cash from a successful ore run, Chalmers was in serious hot water with his workmen. "If Exchequer don't pay in July, there will be fun here," he wrote home. "I have a d[amne]d bad crew."[75] A few days later, he wrote urging Syme, "[F]or goodness sake, try and pay up in July as promised. It's hell now, but it will be something worse if money is not then forthcoming."[76]

Chalmers was not exaggerating the seriousness of the risk. In a letter to daughter Laura, he shared more troubling details:

> The miners who were not paid were very angry, and threatened to put a rope around my neck, and burn the mill and Hoisting Works. For two nights I haven't taken my clothes off, [and traveled] with two men, all armed to the teeth. Thugs are now quiet, as the men have assurance of being paid next week.[77]

In the midst of all this turmoil, Mrs. Laughton had traipsed off to the Palace Hotel in San Francisco, ostensibly to seek medical treatment again for her little son. More to the point, however, she and Chalmers had had a spat about Antoinette's flirtatious behavior and her friendship with a woman in Silver Mountain City of whom Chalmers disapproved.

The nature of Chalmers's relationship with Antoinette had by now clearly evolved beyond that of employer/housekeeper, as his lengthy letter to her in mid-June demonstrates:

> My own dear Antoinette,
>
> Under cover to the [hotel] proprietor … I send you cheque for $40, all I can spare at present…. I was glad to find there was nothing very serious the matter with the little boy. For goodness sake, Nettie, take care of yourself, if not on my account, on your own and Henry's. You must not blame me now, if I am a little distrustful. God knows, I did not wish to feel any[thing] other than implicitly trustful. I am uneasy all the time and can't help feeling so.
>
> I am not going to conceal from you that I am very lonely without you, and the evenings long, and that I would be very, very glad to have you back … [B]ut spare me the anguish of a bitter parting, if not coming back; … only write me that you are happy and let me know where I shall send [your] things. Nor do I want you to come back if you are ever to be at Silver Mountain again. There are nasty rumors afloat (altho' less than I expected) that can only be lived down by not going around or recognizing, in any way, you know whom.
>
> I entertain, you know, a very poor opinion of your female friend there, and think that, but for her, certain things would not have happened. I may judge her wrong, but can't help feeling so, and cautioning you against her…. If you are not to accede to my wishes in this respect, remain in San Francisco or where you choose. How much, let me know, will keep you. God knows, parted as we shall be, I shall do more and all that I can well afford….
>
> Now, my dear Net, be frank with me; any intention of returning to San Francisco soon to take up your new abode there, say so, or if you have not made up your mind to shun San Francisco or Silver Mountain, say so—and we will now, ere anything else happens to you, say good-bye and part good friends. The past is gone, but a renewal of it I could not stand.
>
> Adieu, my dear little Nettie, and believe me, now as ever,
> Your true friend and yours most affectionately,
> Lewis[78]

Nettie returned to Silver Creek a few days later, but not for long.[79] The following month, she was off again, this time ostensibly seeking medical advice for herself in San Francisco, and without advising Chalmers where she had landed. He was obliged to write her doctor to try to discover her address.[80]

The Arnots were still seeking payment, and prospects for financial relief looked grim. "Everyone [in London] feels sore and are doing their best to ruin me," Chalmers complained to Syme. "I *am* ruined."[81] Facing the very real possibility he might soon lose his position, Chalmers penned a letter in late July to Adolph Sutro, who had visited the Exchequer Mill in the past, offering his services:

> Since I had the pleasure of seeing you here I find, from the report of the shareholders meeting in the *Mining Journal*, that much dissatisfaction was expressed with my management—the old story: a mine failure, a bad manager.

> Although I think the directors would like me to remain, I do not like staying where I am not appreciated, conscious as I am that I have worked body and soul for them. Further, my remaining might militate against their getting in additional capital. Being a large shareholder, myself, I must be no barrier in this respect. But if I do leave, I must get something as soon as possible, as every cent I have is in the Exchequer Mine and they owe me a cash balance of over $30,000, which they may not, at once, find it convenient to pay and I do not wish to harass them.
>
> I claim to be a competent, economical, and energetic mine and mill manager, a fine Assayer, and Bookkeeper. Can do a little surveying and am not a bad correspondent. I was bred to the law in Scotland and know something of banking. My antecedents up to the time of leaving Scotland you will gather from enclosed copy letters sent me at the time.
>
> For eight years my time has been devoted exclusively to the management of mines, the erection of the mills and hoisting works. The work here and at the I.X.L. and Imperial Mines will speak for itself. I planned and built the Mill here, through which I showed you. So also the Hoisting Works at the Exchequer and I.X.L. mines.
>
> Should you want a Secretary or confidential agent or mining manager, I am at your service at a reasonable salary. I have a good connection in London. Earl Poulett and Henry Syme, London, are the Chairman and Vice Chairman [of the] directors, and to them I refer you.[82]

Sheriff's sales to satisfy Arnot and other creditors were scheduled time after time, but repeatedly postponed.[83] It would be over a year before the liens were finally resolved in August 1878. But eventually they were, in typical Chalmers fashion: through a roundabout conveyance vesting title in a newly reorganized Exchequer company.[84]

In the meantime, as the summer of 1877 wore on, Chalmers's hands remained tied: "Unless I strike ore in the 200 [level], I stop driving on the 31st [July]," he advised Syme. "[T]he Mill is locked up in the Sheriff's hands, all my quicksilver, salt, tools removed." Crews were continuing to work, but there was no good ore in sight. Still, his optimism was unchanged: "My faith in I.X.L. and Exchequer remain unshaken and considerably bettered," he proclaimed.[85]

Chalmers himself was owed great deal of money, both to reimburse funds he'd had to borrow and as unpaid salary. "The company owes me now $32,200, which I am afraid I shall never see," he confided to friend Rickard. "If so, I have to leave—my family and self nowhere. Still they say 'Never die.' Things may take a more favorable turn. Let us both hope for better times."[86]

Nettie remained under the doctor's care in San Francisco, she said, and was still feeling "poorly." Naturally, she also needed money. Although Chalmers declared himself "impecunious," he induced a friend to send her $100. He was much put out, however, that she still had not provided an address, forcing him to write her simply care of "Post Office, San Francisco."[87]

Chalmers shared his latest spate of bitter news with her: the company was sending yet another mining expert, Henry Sewell, to inspect his operation and "find pay ore that I have passed by."[88] Rumor had it, he added, that Arnot and a friend were on their way to Carson to intercept Sewell, "in order to prejudice him against me, with the view of getting me turned out."

> I tell you this in case you may, in these circumstances, think it scarcely worthwhile sticking to me any longer. All I can say is, that though you may find an abler one, you may find some difficulty in finding a truer friend, and as long as I have a crust, you can share it.[89]

Despite his spectacular failure with the last ore run, Chalmers retained at least one staunch supporter back in London. To friend W. G. Romaine, Chalmers confessed what had gone wrong: "I have learned a lesson by sad experience, and that is, in a spotty mine, to put no faith in assays of samples," he confessed. "Where I erred was in having so much [ore] before putting it through the battery."[90] Chalmers had evidently tested only a few battery samples, whose results led him to expect a modest return of $20 per ton. In reality, however, the total mill run had yielded just $4 per ton.[91]

As usual, Chalmers found a way to put a positive spin on this awful result. The sad outcome at least confirmed his own honesty, he argued; after all, he had not rushed to sell his own shares before the word got out, even knowing the outcome would be dismal.

> Of course I, and I only, knew the ore wouldn't pay, the moment I assayed the first battery samples. But honor prevented me from selling [my shares], although a telegram would have done it, before I sent the bad news officially. I mention these things to you only to prove that, although a fool perhaps, but L.C. [is] no rogue.[92]

Weeks later, Nettie was still in San Francisco for "medical treatment," and Chalmers was getting anxious. "I was terribly disappointed last night when the stage drove past without coming in," he wrote to her on September 4. "I am in hopes however you are only fooling me so as to get more money, and that you are not quite so bad as you say." He urged her not to travel until she was well again, but assured his one-time housekeeper that "when you return, you have no work to do so long as I can afford to pay a Chinaman ... Hasten home, and let me nurse my wee pettie well and strong again, and then, no more bad girlies [*sic.*]."[93]

Chalmers began casting his eye about for fresh projects that might draw in investors, eventually deciding on a Sutro-like tunnel for the Isabelle. "My plan hinted to you of a tunnel to try the whole east side of the Canyon," he wrote Syme. "Look what could be saved in hauling at Exchequer. This tunnel would cut the loss." Though still worried by creditors, especially Arnot, Sr., whom he called a "d[amne]d scoundrel," Chalmers urged Syme to "Keep up your spunk."[94] With friend Wheeler, Chalmers was more candid: "The last failure has gone, nearly taking the gut out of me."[95]

Rats to the Rescue

Sewell, the expert sent by the London board, finally arrived at Exchequer Mill in mid-September for what was supposed to be a three-month visit. The directors were hopeful that his disinterested report, if positive, would help secure fresh capital.

Sewell did manage to placate Arnot, Sr., traveling to San Francisco to meet him and sweet-talking the elder Arnot into accepting half his lien in cash, the rest presumably in stock.[96] However, when it came to reporting on the mines, it soon became apparent that Sewell had his own ulterior motives. "I can only come to one conclusion and that is, that he wants my place for himself," fretted Chalmers in a letter to Syme marked "strictly confidential":

> Yesterday he told me that he was sure his report would be useless so long as I remained at the Mill, and that I ought to ask at once by Cable for leave of absence for six months, and go away, leaving him here for that time as manager.

Sewell irritated Chalmers further by calling him "a stubborn Scot" to his face, and telling Mrs. Laughton that the shareholders detested Chalmers and his own sisters would not write to him. "A more unbearable, egotistical, selfish man I never met," Chalmers fumed.[97] It did not help when Chalmers learned that Sewell had also been "blackguarding me in Markleeville" to anyone who would listen.[98] Ultimately, it was rats that solved the problem:

> This morning [Sewell] came down in a towering rage. The rats made such a noise above his head he couldn't sleep, and the moment he had finished inspection of the mine, he would be off to Carson.[99]

To Chalmers's delight, Sewell departed after only twenty-three days instead of the expected three months, withdrawing to San Francisco where he settled in at the Palace Hotel—at the company's expense, of course.[100]

Sewell's report, when it finally arrived, was nominally favorable—but only if operations were continued under his own management. The notion of working under Sewell was more than Chalmers could stomach, however. "Make up your mind to get rid of Sewell or me," Chalmers chafed to Syme.[101]

Antoinette was finally home again, though still suffering "violent headaches." Chalmers dispatched a note to her doctor in San Francisco requesting a remedy, explaining: "At nights she can't sleep and her memory is affected and [she] fancies disagreeable things."[102]

Back in London, son Johnny had luckily been apprenticed out, making one less mouth for sister-in-law Eliza to feed. But despite Chalmers's insistence that she secure cheaper quarters, Eliza had instead renewed the Ealing house lease for another year. Daughter Ellen, she wrote him, was in need of "sea air," a plan that Chalmers pronounced "an excuse for traveling and spending money."[103]

Closer to home, Chalmers's sometime-friend, sometime-enemy N. D. Arnot, Jr., had settled in at Silver Mountain City, though the widower wouldn't remain single for long. Arnot married Anna Ford, daughter of a prominent hotel-keeper, in October 1877.[104] By December, he had opened his own law office at Silver Mountain.[105]

A £500 remittance arrived from London in December, enough for Chalmers to pay his engineers "to keep them quiet" and lay in winter supplies.[106] Winter snows had already begun, but Chalmers managed to continue preliminary survey work for his planned Isabelle Tunnel.[107]

Anna Ford Arnot, *circa* 1877. (*Alpine County Historical Society, Marcia Bernard Collection*)

Chalmers spent a quiet Christmas at home with Antoinette, though it must have been an anxious time. Just three days later, on December 28, 1877, Chalmers left for the long journey back to London, hoping his deft personal touch could soothe the investors and induce them to part with additional funds for the cash-strapped mines.[108]

By January 15, 1878, he was back in London, penning a letter to Lord Poulett to notify him that "I am here to meet the Shareholders and give an account of my Stewardship—unfortunate though its results have been."[109, 110]

It would be another five long months before he would return to Alpine County—and Antoinette.[111] But when he did finally return in June 1878, Chalmers was more determined than ever to push his ambitious Isabelle Tunnel project.

A tunnel could intersect multiple lodes as it progressed, he reasoned, and would tap the lodes at depth, where the ore (he thought) would prove richer. But the Isabelle Tunnel was far more than just a grand way to economically retrieve ore. It would also be the key, Chalmers firmly believed, to resuscitating the flagging fortunes of the Exchequer and I.X.L.

8

Birth of the Isabelle (1878)

Chalmers allowed himself a single precious night with his family at Ealing after his three-week journey "across the pond." It was a bittersweet homecoming that reminded him just how long he had been gone. As Chalmers quipped afterward: "My children are so grown that I scarcely knew them, and make me quite an old boy."[1]

Yet on January 16, 1878, after one night's rest, he forced himself to push on. His next stop was Paris and then, a few days later, the South of France, paying personal calls on some of the companies' most prominent investors. At last, after more whirlwind meetings on the continent, Chalmers returned to London. But his glad-handing had only begun. He continued to put his natural charm to good use, holding "longish" interviews with more of the firm's wealthy and influential shareholders in London.[2]

Just what verbal magic Chalmers employed in these meetings is not recorded, but undoubtedly, he puffed the potential of the Alpine mines, despite their poor showing in his latest mill run. About this same time, he wrote to an "old Imperial friend," blithely assuring him that the Exchequer "only wants depth to be a second Comstock." The disappointing mill run, in other words, had nothing to do with the true value of the mines; burrowing deeper was the secret to unlocking their riches. It was a refrain Chalmers would repeat in letters in coming months and was almost certainly the same message he conveyed in person.[3]

Chalmers enjoyed a surprisingly warm reception in his face-to-face meetings. "Can't say yet how things may be," he wrote N.D. Arnot, Sr., "but all those whom I have met yet are glad to see me."[4] To N. D. Arnot, Jr., however, Chalmers confided: "but whether they will put their hands deep in their pockets yet remains to be seen."[5] He was wise to be skeptical, as it turned out. Despite cozy meetings with "nearly all the principal Shareholders, all anxious to go ahead," few would prove anxious to throw any more money at their thus-far-unproductive mines.[6]

Syme was in no position to help at the moment. Like many wealthy Brits, Syme and his wife were spending the winter in the South of France. They were not simply chasing sunshine, however; Syme hoped the warmer climate would improve his wife's health and was not anxious to return to London.[7] But his absence left all the politicking on Chalmers's shoulders.

"Return—the sooner the better," Chalmers urged in mid-February. "I am a d[amne]d bad diplomat."[8] Four days later, he wrote Syme again, pressing him once more to come home—"You are really wanted here."[9] Syme was still vacationing a week later, however—now in Naples, Italy. "Unless the waters are stirred soon, we shall all be in a mess," Chalmers growled in frustration. "Come home, come home, come home."[10] A directors' meeting had to be postponed "until H[enry] S[yme] is present," Chalmers chided, because "you are the only one who can manage them—because everyone has confidence in you."[11]

Chalmers penned a gloomy update to Antoinette:

> Things are not looking so well here as I could wish, and I shall not return so soon as I expected, although return I will, whatever happen—but don't say so, to anyone; on the contrary, say you are beginning to fear I won't and are looking out for something else. Make things as black as you possibly can.... What would it cost you to go to San Francisco for three or four weeks and live in lodgings? Write me. If I cable you $200 or so, lock up the house, give the key to Thomson, go to San Francisco and send me your address, and write me from there. Say you have left for good. Trust to me."[12]

The actual situation was currently gloomy enough without embellishment. The Arnots' foreclosure sale of the mines had been postponed yet again, but the newspaper predicted the mines could soon "fall into the hands of a San Francisco company"—*i.e.*, Arnots' Vulcan Iron Works.[13] Meanwhile, the personal property was auctioned off in March to satisfy other creditors, "to good advantage—to the purchasers," as the newspaper tartly put it.[14] Even the long-dormant Imperial in Bulliona appeared at risk. Fearful that Exchequer creditors might try to attach Imperial, Chalmers got John Sauquet, a friendly Silver Mountain merchant, to file a fictitious lien to make it a less-attractive target.

Chalmers's personal possessions, too, were showing up in the paper in delinquent tax notices, with his library, furniture, survey instruments, firearms, and watch all at risk.[15] He had had no money to spare for taxes before leaving for London; he only managed the cost of the trip home thanks to the "quiet services" of the Arnots (a loan) and by dipping into company funds.[16]

On the positive side, however, the Arnots were being more than agreeable. They not only repeatedly delayed their foreclosure sale, but also quietly helped the company buy-out other creditors at discount prices. Eliminating those debts for pennies on the dollar prevented other sales and made it far more likely that the mines could eventually be revived.[17] Since the Arnots had their own large claim still to settle, that was good for the Arnots, too.

London happily participated in this quiet collusion. "You may rely on perfect accord and more from this side," Chalmers assured Arnot Jr.[18] Translation: London would forward funds to buy out discouraged creditors, after which the Arnots' foreclosure could go forward—with title eventually reverting to a reorganized company.[19] All in all, it was a doable if not exactly up-and-up plan to lift the massive burden of debt crushing the two mining companies and resurrect them as fresh British entities.

Restoring their attractiveness as investments was another thing entirely. Not only had Chalmers's recent mill run been disastrous, but Sewell's report had cast doubt on Chalmers's managership.[20] A fresh plan was needed to overcome those objections, and Chalmers had the perfect idea in mind: the Isabelle Tunnel. As he colorfully put it to Syme:

> Tubs that have no bottoms cannot stand on them, and Exchequer and I.X.L. bottoms have fallen out, and will only be resuscitated by working the tub that has some bottom ... [I believe] it will be next to impossible to carry I.X.L. and Exchequer without Isabelle aid.[21]

Crazy on the Isabelle

A tunnel offered the perfect answer to the mines' current "want of depth." It was also multi-purpose, tapping not only Isabelle's own set of lodes but the I.X.L.'s and Exchequer's as well, and offering an easy way to run out the ore—all without the expense of a vertical shaft, pumps, or hoisting. Best of all, at least as Chalmers spun the plan, results would be fast: "In seventeen months you will be taking out Isabelle ore and raising the sparkle of both I.X.L. and Exchequer shareholders with Isabelle Dividends," Chalmers cheerfully predicted. "Four months more would cut I.X.L., and another four months, Exchequer."[22]

To pay for the tunnel, Chalmers concocted an elaborate financing scheme. First, I.X.L. would issue new debentures. Purchasers would get one free, fully-paid-up share of stock in the Isabelle Tunnel for every two debentures purchased, entitling them (in theory) to begin receiving dividends as soon as Isabelle reached her lode.[23] All three companies would share the cost of the driving the tunnel past the Isabelle lode, with Exchequer and I.X.L. paying £10,000 apiece for a one-third right to use the tunnel.[24] "This would ... give you a better chance to get the rest of the money for a live horse, rather than three dead ones," he urged Syme.[25]

There was far more paper-pushing than substance in the arrangement. Its success depended on a great many 'ifs,' not least of which was whether any of the three mines would, in fact, turn out to contain silver. Yet as creative financing plans go, it was an attractive one. Chalmers soon was able to report that at least a few of the original investors were "crazy on the Isabelle project."

Chalmers already had experience with tunnel work, having pushed the sadly barren Imperial more than 1,400 feet into Mount America. That had been an agonizingly slow process, using hand drilling in hard porphyry. This time, he was determined to emulate the success of Adolph Sutro, whose crews had made three hundred feet of tunnel a month "in hardish bedrock" using Ingersoll drills and Burleigh compressors. In an effort to educate himself about this new method of drilling, Chalmers wrote again to Sutro, asking for an "authentic account of the cost and progress in the Sutro, [and] also of the drills used, etc." He did not hesitate to drop Sutro's name and driving rate as he promoted his own tunnel plan.[26]

Though forced to acknowledge his lack of financial success thus far, Chalmers again stressed the Alpine mines' potential. The Exchequer was "a true fissure Lode," he assured shareholders, "carrying stones of ore worth from $100 to $4,000 per ton." Thus far, they had only reached the 400-foot level, he argued, while at depth "the ore will be less broken, more uniform and ... of a good paying average." The problem with I.X.L. had simply been water in the engine shaft with no capital for pumps.[27] His sampling was not to blame for that incredibly disappointing mill, either. "The ore is of a nature that you cannot sample it," Chalmers assured a critical investor.[28] As for erecting a costly mill and other improvements before results justified such expenditures, he contended:

> I do not think I am altogether responsible for this. In any case they would have had to be built someday. When you reach pay ore, and which I think depth will bring you, you will save twelve months' time it would take to build them.[29]

Things looked grim at first for the new Isabelle gamble. On March 2, Chalmers wrote Syme dejectedly: "Nobody here cares a d[am]n whether your mines are lost or not."[30] With Syme still out of town, Chalmers took the liberty of importuning Lord Poulett, urging him to pledge £1,000 to the new venture—or even half that—to demonstrate that the original directors still had faith in the mines.[31]

Back home, miners were pressing for payment and a separate creditor's sale had been scheduled for March 15. Arnot warned in a terse cablegram: "Matters bad." With some $5,000 needed to stave off this latest disaster, the board convened a hurried meeting to authorize the expense. The money was wired to Arnot on the evening of March 14, a hair's-breadth before the deadline.[32]

Nettie's Near-Miss Nuptials

Antoinette's situation, too, was growing increasingly precarious. "My housekeeper threatens to leave, her money is gone," Chalmers fretted to Syme. "She thinks I am not able to stem the storm, and that I will leave her in the lurch and not return."[33] Chalmers did the best he could from a distance, asking Arnot, Sr., to send Antoinette $50 immediately for living expenses and to pay her $40 monthly salary until Chalmers's return.[34]

Yet by late March, Chalmers was losing sleep. On top of everything else, he was in danger of losing Nettie, who was looking to a quick marriage for salvation.

> I am much concerned to find, for the first time, that you are left at the Mill all alone—exposed to the addresses of your old persecutors, and anyone else who may come along.... [A]ny life is better than that of an ill-assorted marriage, as you already know to your cost. You are only a girl yet—there is no need for you to take the first one who offers.... If you have made any rash engagement, break it off ... I will succeed here, though slowly, [and] when I return I will be better able to aid you ... But you must on no account let anyone know that I intend to return, as it is only in this way I can obtain a settlement of the Company's debt for half.[35]

Things rapidly went from bad to worse. Less than a week later, Chalmers received another distressing letter from Antoinette, this one written from Silver Mountain:

> I have been turned out of the house at the Mill, and what I shall do, God only knows. I have no money, and am at Mr. and Mrs. Arnot's. If I had money I would go to San Francisco to live until you come back. I have been on the sick list, too—for I have been

> worked to death. Bemis, your Sawyer, has done it all [turning me out]. The furniture has been moved to Silver Mountain by Mr. Arnot. Little did I think when you left that I would be left without a home, but I hope you will be home soon—else I shall have to find a home some place else. Hoping you are all well at home, I am yours truly,
> A. Laughton[36]

Chalmers dashed off a hasty response:

> Your letter of 14th [March] has given me great pain, but ere this reaches you, Mr. Arnot Sr. will have sent you $50 to take you to San Francisco, and you will be there waiting for me.... I shall never forgive myself if you throw yourself away, before I get back.... Go to the Kellys in San Francisco till then, that I may know where to find you. I will take you from there whenever I arrive. In meantime, I want you to look out for a quiet lodging, two bedrooms and a parlor in a retired place where we can go until I have all my arrangements perfected. I will let you know by telegraph to Mrs. Kelly's the day I expect to arrive, when you can come and meet me at the boat and let me know where to go.
>
> Should you be engaged to be married (which, God forbid), put it off till I see you.... Do not write me here again, as I shall be on my way to California. Burn my letters.[37]

Chalmers's fears that Antoinette would "rush into a hasty marriage" were well-founded.[38] Despite his assurances, she became engaged to Silver Mountain merchant W. L. Flippen, who was operating a store on Main Street in partnership with Arnot. Flippen apparently knew just what to say to win Antoinette's affections. He claimed his brother was the mayor of Memphis and said he was going to be wealthy. She wrote him love letters, and gave him a locket as a token of her affection.[39] What Flippen had not bothered to share was that he was already married and had a family.[40]

It proved a short-lived engagement, however. On April 15, Chalmers received a terse telegram from Antoinette: "Ill. Send money." It was her usual excuse to wheedle traveling expenses to San Francisco. Somehow, Chalmers managed to send the necessary funds, and four days later, received word that Antoinette was safely back in San Francisco with her friends, Isabella and John J. Kelly.[41]

Antoinette had a place to stay, but that did not assure her of a future. She wrote asking pointedly what Chalmers intended to do for her. He was not about to suggest marriage, though he did promise vaguely to "see after" her.[42] He signed his next letter, "With kindest love."[43]

Flippen wound up taking his own life after a dispute with the local stage driver. Whether that had anything to do with Antoinette leaving town we can only speculate, but Flippen loosed a volley of buckshot as the driver was going by. Luckily, the driver survived. However, thinking that he had killed the man, Flippen took refuge in a vacant house, where he shot himself in the temple with his pistol.[44]

Back in London, Chalmers and Syme had finally raised over £5,000 toward the Isabelle Tunnel project.[45] It still was not enough, but the odds of getting the rest were looking better and Chalmers was growing increasingly anxious to be back. He was worried about Antoinette, of course. He also had learned to his dismay that his furniture, books, and papers had been taken from the house by Arnot, Jr., and Chalmers was eager to reclaim them.[46]

In early April, Chalmers dropped a quiet letter to Arnot, Sr., advising that London would soon be sending another $5,000 or $6,000 to aid Arnot's efforts, after which Chalmers would "start for San Francisco as quietly and secretly as possible." He could not afford to show his face at the mine, of course, while Arnot was trying to bargain with creditors.[47] So he had resolved to "travel incognito" to San Francisco, then "go into quiet lodgings" until the liens were all resolved.[48]

Yet Chalmers could not leave London quite yet. He was again out of cash, for one thing, and complained of being "dunned daily" for his family's bills at Ealing. "If I had £300 or so, the rest could follow," he urged Syme in March.[49] But the weeks dragged by without assistance. "Unless something is done here at once, I shall sell all my Exchequer and I.X.L. interests for what they will bring, and leave," he pressed on April 15. "Any life would be preferable to this."[50]

Returning, Though Not Yet Home Again

At last, Chalmers was able to plan his departure. He wrote Antoinette, admonishing her not to meet him at the wharf "in case you might be followed." Instead, he promised to come find her at the Kellys's.[51]

Antoinette had not quite gotten over her pique at being evicted in Chalmers's absence. He smothered her with apologies: "I have never forgiven myself for leaving you to the tender mercies of Alpiners." He closed with assurance that things would be good again: "Trust me, Nettie, as of old. You will not repent it."[52]

He urged Antoinette to look for a furnished apartment in San Francisco that they could hire by the week, "one parlor and two bedrooms, in a quiet but respectable place, not too dear," or a "furnished cottage in the country." Or perhaps, he speculated in a subsequent letter, "Oakland might be a better place."[53] In any event, he stressed, "I want an out-of-the-way place for a little."

By June 1, 1878, Chalmers was safely back in California after his stressful five-month sojourn overseas.[54] Even his trip home had not gone smoothly. While crossing the plains on the Union Pacific, Chalmers had been held up by train robbers who relieved him of his father's gold watch and chain and $100 in cash.[55] Surprisingly enough, not only did the authorities recover the watch and most of his cash, but Chalmers eventually had the satisfaction of learning that the robbers had been caught and sent to prison.[56]

Chalmers kept his promise to Antoinette, and they settled "in the suburbs" of San Francisco.[57] "I do not yet go out in the street, but see Arnot almost daily," he advised Syme, telling him to address future letters "care of Arnot."[58] Even Arnot did not know exactly where Chalmers was now living.[59]

Chalmers was wise to keep his location a secret. Back in Silver Mountain, the sentiment toward him remained bitter and even "hostile," and the risk from disgruntled creditors was elevated.[60] "I hear they have threatened to hang Arnot and L[ewis] C[halmers] if they thought there was any collusion between us to cheat or swindle the Creditors," he confided to son Willie.[61]

Although many creditors' claims had been settled, the debt to the Arnots remained unresolved. The I.X.L. Mill had turned into an unhappy project all around, with Arnot, Sr., likely losing money on the contract.[62] Chalmers was eager to wrap up this last important obligation, proposing at first to pay the Arnots with a combination of cash and debentures. The canny Arnot, Sr., would not agree, at first. But in early July, with his health beginning to deteriorate, Arnot agreed to take a flat $9,000 payment "to cover his whole claim against Exchequer and I.X.L."[63]

The deal was struck none too soon. The senior Arnot passed away on August 27, 1878, "of that fell destroyer, consumption," just one day after the Vulcan's long-delayed foreclosure sale finally took place. As her husband's assignee, Mrs. Arnot became the legal owner of the Exchequer Mill, the I.X.L. Mill, and the I.X.L. Mine. Pursuant to the $9,000 side agreement, she promptly conveyed it all to a trusted agent for the London consortium.[64]

A few loose financial ends still dangled. The Exchequer Mine still had its own foreclosure sale scheduled for November, for instance, and claims against the Isabelle had yet to be resolved.[65] However, Chalmers was confident that those could be arranged, using Arnot Jr. as a middleman.[66] In a letter to an acquaintance, Chalmers boasted, "Now that old man Arnot is dead, I can manage his son."[67]

Eager though he was to begin work on his new tunnel, Chalmers could not afford to show his face in Alpine until the titles were all clear. So for the time being, he was stuck in San Francisco.[68]

While Chalmers was tapping his toes with impatience, the Sutro Tunnel in Virginia City notched its first major win, breaking through to the Savage Mine on July 8, 1878, and connecting with the South shaft two months later. These well-publicized successes further fueled Chalmers's enthusiasm.[69] But while Chalmers hoped to match Sutro's blazingly-fast progress of 300 feet of tunnel a month, he also hoped to cut his workmen's wages to just $3 a day—a move he acknowledged "will scarcely increase my popularity."[70]

He argued with London over the best compressed-air drill to use in the Isabelle tunnel. Rebuffing their suggestion of a Penrice drill, he lobbied instead for an Ingersoll drill, the brand Sutro was using.[71] He even wrote Sutro in August 1878, asking hopefully if he was planning to dispose of any used Ingersoll equipment.[72]

As always, Chalmers kept his eyes peeled for fresh investment opportunities for his friends in London. The Leviathan might still be available, he advised Syme, now for as little as $1,000.[73] The celebrated Tarshish, too, was for sale at a bargain price; both the mine and its twenty-stamp mill could be had for just $25,000; the mill alone had cost $75,000 to build.[74] Chalmers even sent home a photo of the Tarshish Mill, mine, and assay house to help prompt interest.[75]

In local mining circles, however, the real excitement was at Bodie. Chalmers assured his backers that he, too, was on the lookout for "something good" there that could be floated in Paris.[76] Situated roughly 45 miles from Bridgeport, California, Bodie had begun to mushroom in 1876, after large deposits of gold and silver were found. By May 1878, the Silver Mountain *Chronicle* complained that "Bodie has made inroads upon our voters."[77] By August and September, a veritable "stampede" was taking place with residents headed for Bodie.[78]

Relocating Alpiners moved not only themselves and a few trunks; in some cases, they took entire buildings with them as well. A hotel in Monitor and another near Markleeville were disassembled and moved to find new life in Bodie. Butcher R. K. Love sold his two meat markets and drove his band of hogs to the new boomtown. By winter 1878, even the editors of the *Chronicle* had followed suit, hauling

The *Alpine Chronicle* in its original Silver Mountain City location on Main Street near Second Street, about 1867. By 1878, when the *Chronicle* threw in the towel and departed for Bodie, its offices had been moved to the corner of Fifth St. There is no known photo of it in that later location. (*Alpine County Historical Society*)

their printing press and cases over the mountains to re-launch their newspaper in Bodie.[79]

Chalmers, meanwhile, was impatiently cooling his heels in San Francisco, waiting for cash to arrive. He had been "fully assured" when he left London that money would follow soon after his arrival. But more than three months later, the promised funds still had not materialized.

"They write me yesterday from London that on or about the tenth [of] October they will have all three mines floated, certainly the Isabelle, and send me money," Chalmers assured his daughter Laura in mid-September.[80] But he had heard such promises before.

"I shall be looking anxiously for your Telegram after the 10th, by which time, if not before, I shall be penniless here," he nudged Syme.[81] Even if financing for Isabelle was forthcoming, there was still a long road ahead. "Two years is a long time to wait for mine developments," he cautioned.[82]

His recent visit home had apparently led Chalmers to wax sentimental over family matters. Anticipating cash on the way, he wrote an old acquaintance in Fraserburgh, inquiring about the cost of "thoroughly restoring the family burying place, cleaning monument, regilding lettering," etc.[83]

His family was better settled at last. The two older boys, Alick and Willie, were twenty-four and almost twenty-two and "keep themselves now."[84] Willie was rooming with a family named Simons, while Alick had just finished an apprenticeship with a Westminster land-surveying firm and had become, Chalmers boasted, a "first class draftsman."[85]

His three daughters had been sent off to a boarding school in Eichstadt, Bavaria, "because I did not feel justified in keeping up my crib at Ealing."[86] Sister-in-law Eliza had followed the girls there, even though she did not speak German. "I don't think she will stay long," Chalmers sighed.[87]

The much-heralded tenth of October came and went. But two days later, a happy telegram finally arrived: "Isabelle success, Money Thursday." The necessary subscriptions had been taken up; the Isabelle Tunnel project was a go.[88] That same week, an abridged prospectus for the reorganized Exchequer Gold and Silver Mining Company also appeared in the *Aberdeen Journal*, seeking to raise £100,000 in capital.[89]

Money began flowing into the Isabelle's account with Bank of California. A welcome deposit of £2,000 (nearly $10,000) arrived in mid-October, followed by another £500 in November.[90]

Still hunkered down in San Francisco, Chalmers did his best to make good use of his time before the tunnel work could begin in earnest. In late October, he penned a lengthy letter to the prestigious *Mining and Scientific Press*, renewing his plea that foreigners be allowed to patent American mining locations and that non-productive mines be exempt from taxation.[91] He also began ordering the drills and materials needed once the Isabelle started up in earnest.

London was still urging him to try Penrice drills. But Chalmers now had his sights on the National Drill Company's percussion-style drill and was eager to buy two or four of them.[92] These would not come cheap; by Chalmers's estimate, four National No. 2, 3.5-inch drills would cost $2,200, and the two compressors to run them, another $3,500.[93] Add $5,000 for cars, drill steel, sledges, and other miscellany and another $2,000 for piping, and his total estimate for new equipment came to $13,000.[94]

Chalmers remained anxious that the locals at Silver Mountain not get wind of his preparations to start his new tunnel, for fear he would not be able buy off more old debts at a steep discount.[95] Even in San Francisco, caution remained his watchword and he concealed his identity with a pseudonym:

> The other day I was in the Office of Berry and Place, agents for the Drills and Compressors, beating them down on prices, when who should come in but one of our largest unsecured Creditors. Luckily he didn't recognize me. If he had and Berry and Place had told him I was entering into a large Contract for machinery for Silver Mountain, I couldn't have bought [his debt] for less than $1 for $1, and interest and expenses. Luckily also I had taken the precaution not to be L. Chalmers to Messrs. Berry & Place, until we are out of the wood.[96]

He had initially intended to purchase one duplex compressor with a ten-inch cylinder, advertised for $3,000, but that one had already been sold. Instead, the agent offered two single compressors for $3,500:

> [A] 12-inch air and [a] 12-inch water cylinder engine, 14-inch stroke, with a piston speed of 320 a minute, to be run together ... Their [regular] selling price for these was, they said, $2,000 each. I beat them down to $3,000 for the two, and bought them, and have more capacity considerably for my money than I would have had, if I had got the duplex at $3,000.[97]

A boiler would also be needed, and Chalmers had found one in San Francisco: "An A-1 Boiler, 16 feet long and 54 inches in diameter, with forty-five 4-inch tubes, steam drum 36 inches diameter by 6 feet long with breeching smoke stack 30 in. × 40 ft. with bands and guy rods, complete." The only problem was the cost: a pricey $1,185.[98] He also still needed 2,000 cords of wood for the winter.[99]

By early November, Chalmers had secured the services of "an A-1 Foreman" for the tunnel, a man named Bugbee "thoroughly up on all kinds of drills and compressors and machinery, and who has had great experience in Tunnel driving with such machinery." As for miners, he was confident he could hire "all the men I want, already trained to the use of these machines, at $3 per day."[100]

His ambitious Isabelle tunnel would need to run 4,000 to 5,000 feet to reach all three lodes, and Chalmers had set his sights on tunnel dimensions of 7 × 8 feet or 7 × 9 feet.[101] Chalmers estimated that, when complete, the tunnel would cost its backers an astronomical £30,000—over $50 million today.[102] He planned to keep drilling and mucking teams working in swift rotation to match Sutro's running time of 300 feet per month:

> I propose to run three drills on two columns, and have a spare one in case of accidents; bore 6 feet holes before moving drills; blast together with Giant Powder; throw back the dirt, so as to lose no time in running the drills up to the face.[103]

But even in early November, Chalmers dared not return to Alpine—though the directors were hounding him to begin. "You had no right to place me in this nasty position," Chalmers wrote angrily to Syme. "I want to rush things, and it will look as if I didn't.... The poor d[evi]l of a manager always placed in some disagreeable position."[104]

Ten days later, the company was insisting he return to his post though Chalmers saw it as the height of folly:

> I have already pointed out the dangerous position in which you have placed Exchequer and the horribly disagreeable one in which you try to place me. I say try, for I have not quite lost my brains altogether, and have no intention of attempting work up there until I have got in the Exchequer ordinary [unsecured] indebtedness.... This ought to have been, as when I left London I understood it would be, the first thing to be accomplished.... I have never been—in all my troubles with the men, want of money, sleepless anxious nights, surrounded by Molly Maguires plotting for my life and to burn up our mills and hoisting works—I have never been placed in a position which I have felt so disagreeable as that in which you now place me.[105]

Worse yet, word had leaked out that the companies had been refloated with "all the money they wanted," and rumor was that at least one disgruntled creditor was searching for Chalmers's current whereabouts, likely with trouble in mind.[106]

Nevertheless, the directors in London had reached the end of their patience. They dispatched an "insulting" cablegram, threatening to send out another manager if Chalmers did not commence work immediately.[107] Chalmers was sorely tempted to call their bluff, writing "I shall be only too glad to retire." Ultimately, however, he did not send that letter.[108]

Instead, he responded drily: "Your cablegram just now received does not evince the most unbounded confidence in your manager." But he underscored just how much they needed him. "I am afraid ... that a Stranger will find it impossible to lay off the Tunnel, owing to his ignorance of the ground and workings of the three mines," he wrote. Of course, he would not be able to personally show a replacement the ropes, he advised with a certain satisfaction, due to fear of attachments.[109]

On November 30, 1878, the final Arnot foreclosure sale came off at last.[110] Arnot was the nominal purchaser, but Chalmers had quietly supplied him the bid money: $2,000 in Isabelle funds.[111] Chalmers had forwarded the cash to Arnot's Wells Fargo account "as if from [his] mother," depositing it in coin so there would be no record of its source.[112]

The Isabelle Begins

The foreclosure sale finally over, Chalmers could delay his return no longer. It was "now a simple question of bread and butter," and Chalmers knew which side his bread was buttered on. Writing to Arnot, he put the best possible spin on his decision to return:

> [A]s no deputy can lay off that Tunnel as I wish it done, and as, in a question as to the moral effect which the mere running of the Isabelle Tunnel will have on our chances of an easy Settlement with the balance of the Creditors of the Exchequer Company, it appears to me to be immaterial who runs the Tunnel, so long as some of the old Exchequer people are interested therein, I have made up my mind to take the bull by the horns and commence work.[113]

Mrs. Laughton would head for Silver Creek in a few days "to put the house in order," he advised Arnot, asking him to give her "all the assistance necessary to get the sticks of furniture replaced where they came from."[114]

But he was not happy about the decision. "I go at your risk," Chalmers admonished the directors. "For, though this last Sale will have disposed of all the Liens, there are still judgments against the Company, or rather against me as apparent owner, for Company's debts, unsatisfied to the extent of $9,500."[115] Even were the company to cover that amount, it still would not compensate some of the most honorable creditors.

> If only this $9,500 is bought up, ... others are left without a cent. John Sauquet [for example], who ... gave me $500 to get rid of a band of incendiaries who had plotted to burn your works, and hang your foreman and manager, and of whom the Sheriff and County Judge were so much afraid that they refused to accompany me to the Boarding House and drive them out: that man gets nothing, not even his $500.[116]

It was all most provoking. Privately, Chalmers railed to Syme: "[I]f I had what is fairly my own from Exchequer, I would not stay with you ... one day." Nevertheless, he dutifully prepared to depart for Alpine on December 11.[117] "Not a word," he cautioned Mrs. Laughton, who had gone on ahead, "even to [contractor] Ginn or

Arnot."[118] On the appointed day, he dispatched both a quick telegram and a letter to London, advising that the latest title documents had been mailed, adding: "I leave for the Mines tonight ... Arnot is very angry at my going. Writes his mother that I must be mad."[119]

Leaving San Francisco as promised on the evening of Wednesday December 11, Chalmers rode the "Lightning Train" to Carson, where a buggy met him the next morning to carry him back to the Exchequer Mill.[120]

He and Nettie tried to put the house back in order, but Chalmers was irritated to find a few things damaged or missing. Most were simple household items, but four expensive ivory-handled breakfast knives and a wringer-style washing machine also had disappeared.[121]

Chalmers also wrote angrily to J. W. Haines of Genoa, complaining that his favorite riding mare was returned "such a wreck that I scarcely knew her." "She has evidently been in harness, and in harness racked and strained, and is now useless for my purpose," he fretted. "You cannot imagine how vexed I was when I saw my old favorite."[122] Now in need of another horse, Chalmers asked for a "trial" of a black mare from noted horseman Ben Palmer of Sheridan, offering $125 "if she suit [*sic.*]."[123]

On December 14, a cablegram arrived, promising: "[£]500 Saturday, another week."[124] Sure enough, a deposit of $2,412.06 hit the company's Bank of California account on December 16, and a similar deposit arrived on the 21st.[125] That was certainly encouraging, but Chalmers had learned his lesson. He wrote to the Isabelle's managing director: "I wish to know whether I can depend on regular remittances."[126]

Work on the Isabelle could now begin in earnest. Having delayed purchasing drilling equipment until now, Chalmers finally placed an order for two, 12-inch National Drill Company compressors "fitted to run, together with all connections," and two Number Two drills "with wrapped hose fittings, wrenches, clamps and all other necessary appendages and connections, and two drill columns." The total came to $3,850, and he had driven a hard bargain for them indeed—including a guaranteed refund if the rig failed to give "complete satisfaction."[127]

Chalmers had already arranged for surveyor L. L. Hawkins to come up from Berkeley to lay off the tunnel line.[128] But Chalmers was leaving nothing to chance; he planned to double-check Hawkins's survey himself.[129]

The proposed Isabelle tunnel site lay on the far side of a creek, with the road on the other. That meant a bridge across the creek was needed to get machinery to the tunnel site. On December 14, Chalmers signed a contract, and soon eight to ten men were hard at work building a wooden bridge. A new boarding house was also being thrown up near the I.X.L. Mill.[130]

Chalmers proposed to hire Bugbee as foreman for $150 per month, cautioning that he would have to pay board of $1 per day. "At present I have no accommodations for Ladies and in fact rather object to them at the works," he observed, "but I can get you a small house one and a half miles off for your wife if you must have her with you. You would have to be at the works."[131]

He also began scouting for men to do the preliminary grading, offering $2.50 per day, less $1 for board at the company's boarding house. "The wages are low," he admitted candidly, "but [the] Company cannot afford to pay more for that class of work."[132] The reduced wage was "something new in this part of the world," he confessed in a letter to the Isabelle's managing director, but Chalmers was callous

The remains of the Isabelle Boarding House, *circa* 1920. A mound of waste rock taken from the Isabelle Tunnel is still visible on the right. (*Dustman collection*)

about the low wages. "[S]o long as I am in funds with which to pay off a growler, it does not matter to me whether they like it or not."[133]

Chalmers also wrote for a Chinaman to do washing for the miners, explaining:

> There is only one Chinaman in Silver Mountain and he is so unobliging, I will not let him have the washing.... Send me if you can, by next stage, a strong, burly Chinaman who can lick the other one if he interferes.... He must bring blankets, tubs and irons.[134]

Hawkins finished his survey for the tunnel on Christmas Eve.[135] On Christmas Day, Chalmers's new grading crew was hard at work, hacking out a spot near the proposed tunnel's mouth for a building to hold the boilers, air compressors, and blacksmith shop.[136] With local lumber hard to come by, Chalmers economized by purchasing a nearby building from another mining company, tearing it down, and re-erecting it on site to serve his purpose. He also decided to forgo the expensive new boiler, and began looking instead for a pair of second-hand ones to be hauled up the mountain before winter snows made the journey impossible.[137]

By now, Chalmers had firmly settled on his tunnel's dimensions: 8 feet high by 9 feet wide.[138] This was "rather larger than I intended," he acknowledged, but "a foot does not materially increase the expense, while it gives so much more space for work, and these dimensions will be none too large if the Tunnel is to be used by three companies."[139]

He was surely on the brink of success at last. There was money to run the tunnel. The survey was complete. A full crew was on hand. A bright future for the Isabelle lay ahead.

Not everyone shared his optimism, however. The *Carson Valley News* that December branded the Exchequer and Isabelle likely "down-right swindle[s]," notwithstanding their prominent British backers. London's *World* and *Daily News* had trashed the Exchequer as a mere "lottery," the column pointed out. Chalmers came in for his own helping of snark: "[W]hilst here [in London]," the article sniffed, he "certainly did not rise to the surface."[140]

If Chalmers had the misfortune to read those blistering comments, they must have stung indeed. But he would have quickly pushed such remarks aside, confident that the Isabelle would—very soon—prove the nay-sayers wrong.

9

Tunnel Progress and Wedding Bells (1879–1880)

Despite Chalmers's misgivings, a welcome influx of £1,500 arrived in January 1879 to fatten the Isabelle's bank account.[1] Winter weather had arrived, but the project was off to a good start.

The lumber salvaged from a nearby building was speedily transformed into a 74 × 32-foot structure at the mouth of the soon-to-be-commenced tunnel. Its interior was partitioned into four, 32-foot-long compartments: an engine and boiler room, each twenty feet wide; and blacksmith's and carpenter's shops, 16 and 18 feet wide respectively.[2] Recycled lumber was not his only nod to economy; as a cost-cutting measure, Chalmers also opted to provide simple candles for his tunnel workmen in place of more-expensive miner's lamps.[3]

But when it came to his own family, Chalmers finally felt flush enough to share the fruits of his well-padded wallet. In February, he transferred some much-needed cash to daughter Laura in Germany. That bit of largesse was accompanied by a polite request that she knit him some socks, as "Mrs. Laughton is not efficient at darning, but does her best."[4]

On the home front, he was still trying to keep Mrs. Laughton content, ordering a wide variety of plants and seeds for her garden, including strawberries, raspberries, roses and lilies.[5] He also sent off for *Woodward's Country Homes*, a collection of "country" house plans, and a subscription to the *Waverley* magazine, a women's weekly filled with music, poetry, novels and the like.[6] In early March, he even purchased eight shares of "Bodie" stock in Mrs. Laughton's name, though whether those were actually for her or simply intended to keep the stock out of his own name is unclear.[7] For himself, he indulged in new subscriptions to the San Francisco *Daily Chronicle*, the *Mining and Scientific Press*, the *Engineering & Mining Journal*, *Scientific American*, and the *American Machinist*.[8]

Perhaps most significant of all from Antoinette's point of view, Chalmers wrote to San Francisco in late March to request a "servant girl." His requirements were quite specific. The helper was to be:

> hard working, cleanly, respectful, obedient, well-behaved … no gossiper and willing to do as told.… She has to do the washing, ironing, assist in cooking and general housework. She will have a good home but no society, wages $15 per month.[9]

Above left: Water still seeps from the mouth of the Isabelle Tunnel (2008). The tunnel site is located on private property; permission from the owner is required to visit. (*Courtesy of Judy Wickwire*)

Above right: The grave of Antoinette's parents in Grass Valley, California. The inscription reads, "In memory of Bavia Depuy, born Jan. 1790, died July 16, 1865, and of his wife Catharine S. Burger, born Oct. 25 1821, died Dec. 13, 1866; erected by their affectionate daughter Antoinette Depuy, Sept. 1879." (*Dustman photo*)

Mrs. Laughton, it seemed, had not only been relieved of sock-darning duties, but also would no longer have to do much housework. For the men at his Isabelle boarding house, Chalmers laid in foodstuffs that, for once, included a few luxury items, including "Japan tea" and tinned oysters.[10] For himself, he inquired about the price of Havana cigars, specifying Cuban-made goods only, "not New York nor San Francisco."[11]

There was enough money, too, to assist Mrs. Laughton with projects near and dear to her heart. Late that summer, she traveled to Shaws Flat, California, to "see her old home and to have her father and mother's burial place fixed."[12]

Battles with Arnot

By May 1879, Chalmers was already complaining that he had been forced to dip into his own salary to pay "private debts and expenses" incurred on the company's behalf, calling it "pretty rough usage to a faithful manager." Despite the seemingly warm relationship he had developed with the Arnots, Chalmers suddenly was "having trouble with young Arnot."

At first, it had been strangely difficult to get "any account from [Arnot] at all" for all the money that had been sent to pay off creditors.[13] Then, after the requested accounting finally arrived, Chalmers discovered what he considered "dirty tricks" buried in Arnot, Jr.'s figures. Where liens had been purchased at fifty cents on the dollar, for example, Arnot's accounting listed the difference as a debt still due—to himself.[14] When Chalmers contacted two of the creditors that Arnot listed as paid, he discovered they had not actually been paid at all.[15]

Those were not the only things ruffling Chalmers's feathers. "Arnot Jr. is playing traitor to me," he growled to a friend, upset that Arnot had seized I.X.L.'s timber land for a second time and sold it, after Chalmers thought everything between them had been resolved. Chalmers dispatched a complaining letter to Mrs. Arnot about her son's recent actions, a step that likely did little to smooth matters over.[16]

Arnot, Jr., was a "Judas" and "little better than a thief," Chalmers declared angrily, ranting in multiple letters about Arnot "misappropriating money" and doing "fraudulent things," labeling him "the most bare-faced Liar I ever met."[17] "He is a bad egg," Chalmers grumbled to Syme, "but I can checkmate him."[18]

Arnot, for his part, was not above taking steps to embarrass Chalmers. When Chalmers refused to voluntarily accept service of a complaint filed by Ford (Arnot's father-in-law), Arnot bypassed the usual procedure of having the sheriff quietly deliver the summons. Instead, he aired the dispute publicly by publishing the notice to Chalmers in the local newspaper—a paper he knew Chalmers would not receive while in London—and then obtained a default judgment.[19]

As if Chalmers needed more cause to be livid, by June 1879, Arnot had secured the powerful post of recorder for the Silver Mountain mining district. This not only gave him the inside scoop on the latest mining developments but, in practice, also allowed him to decide which mining claims would—and would not—be filed. Arnot's selection as recorder, too, had been obtained by dirty tricks, Chalmers claimed, alleging that Arnot had quietly changed the time of the meeting to ensure few miners would appear, then arranged for a friendly group to show up and vote.[20]

But the thing that perhaps irritated Chalmers the most was the discovery that it was Arnot who had tried to get Mrs. Laughton to marry Flippen—Arnot's partner in the I.X.L. store. After Arnot hinted to Chalmers "that [Antoinette] was not what she ought to be, and that he would write my family," Chalmers hit the roof.[21] He composed a literary retort filled with verbal venom, beginning with a Latin snub, *quod deus vult perdere, prius dementat* ("Those whom God wishes to destroy, he takes away their reason"), and continuing:

> [Y]ou have recourse to tricks to obtain money that no honorable man would lend himself to, ... and a dastardly and unmanly attempt to vilify the character of a woman

> who is as much the superior of her would-be Vilifier as is an honorable woman of a dishonorable man.... Reminding you that ... it is possible for you to allow your slanderous tongue to wag too freely.[22]

Sadly, Arnot's response has not been preserved, but tempers no doubt remained elevated on both sides for some time.

On a positive note, however, by May 1879, Chalmers had finally received the sheriff's deed confirming he officially had title to the old Imperial Tunnel property at Bulliona.[23] At the Isabelle, he also had three shifts of miners working around the clock, pushing his new tunnel forward.[24]

As usual, however, progress was three steps forward and two steps back. The compressors, boilers, and drills Chalmers had purchased from Berry and Place gave him a "very great deal of trouble."[25] On arrival, the drill columns proved too short. Then one of the compressor's cylinder heads proved to be "honeycombed" (defectively cast). When the company finally sent a replacement, it was for the wrong end of the cylinder.[26]

A few weeks later, one of the drill clamps broke and Chalmers had a hard time getting another. As a temporary fix, he had the blacksmith forge a substitute from iron but this, too, soon broke.[27] In desperation, Chalmers traveled all the way to San Francisco to get his hands on a replacement, only to have yet another clamp break soon after his return.[28] Mechanical troubles would continue to dog him into the fall, when a new drill arrived *sans* chuck, feed crank, or clamp, "nor a single wrench which always accompany the drills."[29] Something, it seemed, was always going wrong.

The Ingersoll Drill—the brand Sutro was using. After considerable research, Chalmers opted to purchase three National drills instead. (*Manufacturer and Builder Magazine (July 1879)*)

Chalmers's dispute with Arnot over Exchequer and I.X.L. debts continued, and by July, he was complaining to his lawyer that Arnot (as mining recorder) was hampering Chalmers's ability to inspect the mining records and "leading I.X.L. and Exchequer a merry dance":[30]

> Being anxious to ascertain what Relocations Arnot had made, dates etc., I employed a Mr. George W. Jones last Wednesday to call on Arnot and ask for a look of the Records, which he did in a quiet gentlemanly way. Arnot replied, 'I cannot show you the records, but will furnish you copies of what you wish.' After conversing about half an hour on general topics, Mr. Jones repeated his request and received the same reply.
>
> After a little further conversation on other matters, Mr. Jones a third time asked to see the records, to which Arnot replied that he couldn't show him the records then, but would do so any future time he (Jones) was passing bye. Mr. Jones then requested Arnot to fix a time when he could see them. Arnot stated he was going to Monitor next day, but if Mr. Jones would call at 2 p.m. next day, he would have the opportunity of examining the records.
>
> Yesterday at 2 p.m. or between 2 and 2:30 Mr. Jones called. Mrs. Arnot came to the door. Mr. Jones asked for Mr. Arnot. Mrs. Arnot answered that Mr. Arnot had gone fishing, and had been gone quite a while. Mr. Jones then asked Mrs. Arnot if he would be allowed to look at the Mining Records. Mrs. Arnot said she had nothing to do with them; but invited Mr. Jones into the office, saying in reply to Mr. Jones' enquiry when he [Arnot] would likely return, that he might probably return soon. Mr. Jones remained in the office until 8 p.m. without seeing anyone.
>
> About that time Jones left, but had the curiosity to take an occasional look at the premises as he retired. Saw a female leave the house by the rear and enter the fence enclosing a Barn or Stable on the Summerville Lot some distance from Arnot's house, where he, Arnot, has his Court. Just as the woman left the dwelling house Jones saw a man (too far off to say who) pass from one of these Summerville outbuildings to the Barn aforesaid.
>
> So much for friend Arnot. Now, irrespective altogether of the legality or illegality of his election, his conduct in this matter would appear to me to disqualify him for the office. *Entre nous*, I think he is either manufacturing the records, or he is determined that I shall go personally and see them and him.... Now if we can beat him on this why not commence suit at once. I must have these records in safe hands.[31]

As if that were not bad enough, "Arnot has got the old recorder in Monitor turned out (ten years in office) and put in a creature of his own," Chalmers growled to his foreman, Bugbee. "So he has now the control of both districts."[32] By mid-July 1879, it appeared the only resolution would be "a big law suit with young Arnot."[33] Chalmers hoped that the state Supreme Court "may perhaps take it up as urgent."[34]

Complicating the picture, Arnot was running for local Superior Court judge, with the election coming up in September. Chalmers engaged in a flurry of behind-the-scenes maneuvers to secure a more friendly candidate. "If not too late, send an A-1 honest Superior Judge here, and we will elect him," he assured Bugbee. "I have seen some leading men from the Valley, all [are] favorable."[35] A week later, another outside candidate had appeared on the scene: Rufus E. Chick from Kern County had begun stumping in Markleeville. "They are running a man for this office sent over by Judge Reed from Kern," Chalmers wrote in frustration to Bugbee. "Hurry up our man."[36]

Chalmers was not above a quiet bit of voter bribery if it helped his contender win. He promised one creditor he would pay a bill in full—if his favored candidate was elected.[37] By August, however, it looked as if Chick had the upper hand. That still was preferable, however, to having Arnot on the bench. "I do not trust friend Arnot one bit; he is altogether too mercurial for me," Chalmers confided to Bugbee, adding confidently: "Arnot will be beat by Chick."[38]

Nevertheless, Chalmers hurried to resolve matters with Arnot before the election. He sent off another settlement offer, which he told Bugbee he did not think Arnot would accept, but "I do not see that I can do more in the way of sacrifice."[39] Arnot kept him waiting but, after a personal visit from Chalmers, eventually accepted. Chalmers happily reported that he and Arnot "have smoked the Calumet of peace."[40]

Although Arnot had previously taken the drastic (and, to Chalmers, underhanded) step of relocating Exchequer's mines, he quickly signed over the relocation deeds "and professes a great desire to aid me in settling with Scott etc.," Chalmers crowed.[41] By August 18, 1879, Chalmers was able to mail off a large packet of paperwork to the company lawyer in London, reflecting assignments by Arnot to the company's agent, a man named Dunne.[42]

Fumes and Other Headaches

The Isabelle Tunnel was advancing at a rapid pace. By mid-August 1879, the Isabelle was in 734 feet, and by 8 September 1879, another 204 feet of tunnel had been added.[43]

Chalmers negotiated hard on the price of blasting powder, playing off the Vulcan Powder Co. against competitor Bandmann & Nielsen and extorting deliveries in advance of payment. He even snuck some of Bandmann's Giant Powder "quietly into a Vulcan box," apparently in a covert effort to determine "which of the two is the better powder."[44]

The expense of powder was a big deal. The Isabelle crew was blasting twenty-six to forty holes every twenty-four hours, eating up 100 lb of powder a day.[45] Over a nine-month period, Chalmers estimated his new tunnel consumed $13,500 in powder alone.[46] So competition for his business was naturally strong. Bandmann sent out a Giant Powder representative from Virginia City to pay a special visit to Chalmers.[47] The Vulcan sent their own agent, who "did his best and if any man could, he could," but in the end did not succeed in getting Chalmers to place an order.[48]

Three miners quit in August 1879—partly due to poor pay, but primarily because the fumes from the Giant Powder were so bad. This had been a continuing problem for months; as early as June, Chalmers wrote Bandmann & Nielsen noting that the "men complain greatly of fumes giving more headache than usual," while in July he had described the fumes as so overwhelming that "the men drop down frequently from their effects" and "some men who tried to go in after an hour were struck down as if paralyzed and run out on the stone Car almost lifeless."[49]

Noxious fumes, as Chalmers advised the Isabelle Company, were "the great drawback to all nitroglycerine compounds." The effects could be devastating; he had seen "three and four [men] at a time ... run out on the Cars senseless."[50] A later letter described men "more dead than alive," some tearing "their strong

The brick powder magazine was located a short distance away from the Isabelle Tunnel's mouth, for safety. In 1976, almost a century after it was built, the magazine's brick face was falling away, but the interior was still in good condition. (*Courtesy of Judy Wickwire*)

Miners' blouses in ribbons and with apparently no effort, rip[ping] in pieces the empty Dynamite Cases."[51]

In an effort to mitigate these harsh effects, Chalmers attached an 8-inch, 1,400 rpm Sturtevant blower to a 7.5-inch air vent, and later experimented with a steam jet to help exhaust the gases.[52, 53] He was eventually forced to discontinue using Bandmann & Nielsen's powder altogether "on account of the deadly effect of the fumes on even old Miners."[54]

Chalmers had long suspected the post office at Silver Mountain of opening his letters, and sometimes his mail simply did not arrive at all.[55] He had complained in February 1879 about having to pick up mail at Monitor, "as the Stage does not come up, giving as an excuse the state of the roads."[56] He was especially worried about potential prying by Arnot, who had become the temporary postmaster at Silver Mountain in Ford's absence. Finally, Chalmers took matters into his own hands, proposing that postal authorities open a branch post office at his Silver Creek compound.[57] This request was granted in April or May, with Chalmers's own bookkeeper Thomson pressed into service as postmaster.[58] "[E]rgo, now I have a P.O. here called Silver Creek," Chalmers advised Syme in June.[59]

The arrangement didn't last long; it did not need to. By September 1879, Chalmers was advising his correspondents: "The Post Office here has been discontinued, and Mr. Ford has returned to his post at Silver Mountain. So please address 'Silver Mountain' as before."[60] That would not be the last time that complaints were aired about the local postal system. In 1881, an anonymous letter to the Genoa newspaper complained bitterly about the condition of the Alpine mail, asserting that the route to Silver Mountain had originally been let in 1876 for "about one-fifth of a living rate," then handled by N. D. Arnot as the "leader of the county ring" before being turned over to a "bankrupt teamster from Carson" who did as little as possible, putting his poker games first.[61]

Arnot soon had more important things to tend to than the mail. Much to Chalmers's chagrin, Arnot triumphed over his competition at the General Election on September 3, 1879, winning the post of Alpine County Superior Court Judge.[62] Arnot and his wife quickly moved from Silver Mountain to the county seat at Markleeville. Though Chalmers did not dwell on this development in his letters, he must have realized that, now more than ever, it was important to remain on friendly terms with his sometime-adversary.

By late summer 1879, expenses for the Isabelle tunnel were mounting. The Alpine County assessor, recognizing the tunnel's recent progress, raised the taxable value of the Isabelle property by $21,850.[63] Chalmers did what he could to try to drum up cash, attempting to sell the now-unnecessary hoisting rigs, safety cage and boilers from I.X.L. and Exchequer for about $8,000.[64] But by October, he was pointedly advising London that the current month's expenses were $4,396, and he would need $5,669 more to get through to the end of the year.[65]

Mrs. Laughton paid another extended visit to San Francisco that September. She was gone so long, in fact, that Chalmers's patience began to wear thin. He wrote to her repeatedly, sending small amounts of cash and asking when she would come home. Finally he huffed in exasperation, "If you do [intend to stay all winter], let me know that I may run down to San Francisco, and get another housekeeper."[66]

The threat apparently worked. By October, Nettie had returned to Silver Creek. She had done some expensive shopping while she was away; in November, Chalmers dutifully remitted payment for a piano and stool that Mrs. Laughton had purchased while in San Francisco.[67]

Good news and bad news from Chalmers's family in Europe arrived that fall. Son Alick had secured a position; his father wrote, "I am glad you have got something to do."[68] But in Bavaria, daughter Laura was not feeling well. Chalmers admitted to being "a little anxious about her," especially as "I always looked upon her as strong as a horse."[69] Chalmers himself had caught "a horrid cold" that fall, which he blamed on "traveling [with] an open railway car window," perhaps on a visit to San Francisco where, among other things, he had a photograph taken to send to his children—and to Mrs. Laughton.[70]

His personal bankruptcy in Scotland was still dragging on, much to Chalmers's dismay. "I have heard from Davidson and Gardeau, and what a nice mess they have made of my estate.... Disgustful [*sic.*]. I have heard that Anderson made away with £4,000.... If my creditors have not been paid, it is not my fault."[71]

Challenges continued on the Alpine front as well. Thomson, Chalmers's bookkeeper, departed in September for greener pastures. That was no great loss as far as Chalmers was concerned; "he was always growling about the post office and not being paid," Chalmers wrote home.[72] As a replacement, Chalmers hired George W. Jones—the same gentleman who had shown such doggedness in pursuing Arnot and the mining records.[73]

The Isabelle tunnel now stretched an impressive 1,050 feet.[74] They had struck a "stringer of quartz" assaying $13 per ton in gold—not much, but a promising sign.[75] Chalmers wrote encouragingly to Syme: "I tell you, Syme, we have here property, when opened up, not equaled [by] the Comstock. Keep on running here and I will carry you safely across.[76]

With winter approaching, Chalmers again placed a large supply order. For himself there were lobster, oysters, salmon and sardines, plus "best Japan tea" and "best Carolina rice." For the Boarding House, he stocked up on beans, codfish, macaroni, vermicelli, and cornmeal.[77]

He also ordered furniture-making materials to keep himself occupied during the winter. These included a lathe and other turning equipment plus rasps and a "pin vise." A few weeks later, he wrote son Willie, asking him to send out "Haltstaffle's latest work on turning."[78] One of his most fascinating purchases was "rubber clothing" for himself—an order that included his measurements, affording us a clear image of Chalmers himself. By his own description, he stood 5 feet 9 inches tall and a stocky 50 inches around the shoulders, while his head was just over 22 inches in circumference.[79]

Once again, however, money was tight. If funds failed to arrive soon, he would be forced to stop work. "[I] sent you such a Statement as there is no misunderstanding, and I now await with anxiety a Cablegram from you which will either Continue the work or shut it down on the last day of this month," he wrote the company secretary on November 7.[80] He had at last finally obtained sheriffs' deeds "of the whole [Exchequer] property."[81] But taxes were still due, and in order to get the deeds from Sheriff Scott, Chalmers had been forced to give him "my own personal obligation" for $1,300 in costs.[83] He observed pointedly that there was still the not-exactly-small matter of $29,174.39½ that the company owed Chalmers.[83]

The money, as usual, did not arrive right away, though another £2,000 did appear in December 1879.[84] In the meantime, Chalmers was in penny-pinching mode. He gave his foreman, Bugbee, a hard time for using "altogether too much powder," advising him to "use only half a case at a blast, tamp with water, carefully greasing your primer cap."[85] Before long, Chalmers became so upset with the slow progress in the tunnel that he discharged Bugbee entirely.[86] It was a smart move;

Chalmers enjoyed wood-turning, a popular pastime in the 1870s and 1880s. This advertisement for a turning lathe ran in the *Alpine Chronicle* newspaper on 13 January 1877.

he would later learn that "a good deal of whiskey [had been] consumed about the works at night" under Bugbee's watch.[87]

Desperate for a replacement foreman, Chalmers wrote to Sutro, Vulcan Ironworks, and others for recommendations. Unmarried men were preferred, he stressed, and he was willing to pay up to $7 per day.[88] He eventually hired Robert Andrews, "one of Sutro's best men," for the astronomical sum of $8 per day. "I have a good man, I think," Chalmers assured Syme, "who worked on the first Burleigh Drill set in the Sutro tunnel at Virginia City."[89]

Andrews immediately requested two more drills, an electric battery, electric fuses, and three shifts of workmen.[90] "It appears Sutro spared no expense," Chalmers chafed. "I don't like him so well as Bugbee, but he may make more tunnel, which is everything."[91]

Chalmers continued to tout other nearby mining possibilities, sending friend Wheeler a box of ore from the fabled Tarshish Mine and forwarding "an offer of the Morning Star Mine" to Syme.[92, 93] In late December, he suggested he could pick up the Colorado No. 2 for $75,000 and hinted, "If I had a little spare capital, [I] could do some dead work prospecting."[94]

Antoinette was still causing dust-ups at home. Flora, the housekeeper, quit a few days before Christmas, after Mrs. Laughton "hurt her feelings." Chalmers immediately wrote away for another one, "a good, steady, middle-aged, Scotch or German woman.... An American won't fill the bill, nor Irish"—someone over sixty years old, he specified, perhaps hoping to avoid any jealousy.[95]

Winter had now arrived in earnest, with four feet of snow on the ground.[96] But the year was closing on a hopeful note. The additional £2,000 arrived on December 21, and the Isabelle Tunnel had been pushed 1,600 feet into the mountain, at a cost of roughly $27 per foot.[97]

As December wore on, Chalmers planned a short trip to the Palace Hotel, likely hoping to enjoy Christmas festivities in San Francisco. It is possible Antoinette went with him. If she was hoping for a marriage proposal, however, it did not arrive. Instead, he bought her a "lettered" Bible as a Christmas gift.[98]

Dodging Disaster

The new year brought all the previous year's challenges plus new ones as well. On January 13, 1880, seven men quit. Chalmers's engineer and all four of his drillmen had "struck for wages," he wrote, and both shift bosses left from "sickness caused by fumes."

Unwilling to wait for replacements, Chalmers pressed his bookkeeper, foreman, and carpenter into service in the tunnel.[99] But he knew he had to do something different if he was to keep men. The following week, he departed for San Francisco to procure mining supplies and equipment to improve the air in the tunnel. On his shopping list were a Baker blower and 2,000 feet of 13-inch air pipe, a Laflin & Rand battery, two additional drills, electric "exploders," and parts for a drill carriage, all of which would cost about $4,000.[100]

Fixing the ventilation problem proved not just expensive but time-consuming. Workers had to cut the roof of the tunnel to make room for the large ventilation pipe, locating it high enough be out of the way of the "pull carriage" (tramway). "Most of it requiring hand drilling," Chalmers chafed to the company's secretary, adding "it has been quite a job."[101]

Sheets of galvanized iron, rivets, and shears soon arrived, allowing ventilation pipe to be made on site at a shop near the mouth of the tunnel.[102] His initial 'tinner' was an "illiterate German."[103] But before long, Chalmers decided the man "won't do" and was on the lookout for another—for wages of just $3 per day.[104] His second hire was apparently George Deichmann, another German, age forty-three.[105]

The improvement in working conditions was immediate once the larger ventilation pipe was installed, despite only a small No. 4 Sturtevant blower moving the air.[106] Work in the tunnel began again, and by early March, it reached over 2,000 feet into the mountain.[107]

Chalmers openly touted his tunnel as a "second Sutro." The Isabelle was indeed similar in size, if not in length; the Sutro Tunnel's dimensions were (Chalmers noted) 10 × 8 feet, while the Isabelle was almost as large, at 9 × 8 feet.[108] Eager perhaps for a bit of reflected glory, the canny Scot arranged a personal audience with Sutro during one of the tunnel magnate's visits to San Francisco.[109]

Things seemed to go better financially—at least for a time. For 1879 alone, the Isabelle's expenditures had topped $62,000.[110] On January 19, 1880, Chalmers acknowledged receipt of another £1,500 and the syndicate was promising £1,000 more would arrive soon.[111] To Chalmers's disappointment, however, just £800 ($3,869.07) actually showed up.[112] Company officials dismissed the original promise as a mistake; £800 was all that had actually been intended, they said. That explanation upset Chalmers, who begged them not to repeat such an error; "It gets me in serious trouble here," he chided.[113]

Fresh financial disaster was narrowly averted in March, when sales of the Exchequer and I.X.L. properties for back taxes had again been scheduled. The company cabled the necessary money, but waited to the last second to do so, remitting funds on the very day the sale was scheduled to occur.[114] Again putting his Herculean powers of persuasion to work, "[a]fter a good deal of talking" Chalmers persuaded the sheriff to "deviate from his rule" and immediately hand over certificates of sale and assignments for both rescued properties.[115]

Meanwhile, fresh trouble had erupted in Germany, with sister-in-law Eliza suffering "semi-crazy moods." Eliza was terming her condition "paralysis." Chalmers's daughters politely described it as "a general physical breakdown, brought on by brooding over her monetary matters." Chalmers himself, however, cut through the fluff, writing: "Mrs. Chalmers is insane." He suspected the condition was inherited, observing in a later letter, "She had a brother, Pat, the same."[116]

Chalmers wrote to the school where his girls were boarding, sternly admonishing the administrators not to let his children visit their aunt except for occasional "short visits, and then only accompanied by someone from your school."[117] "A horrid business," he huffed a month later to Syme.[118]

Chalmers's personal bankruptcy was also weighing on his mind. Now in its fifteenth year, the proceeding was still dragging on. On March 5, Chalmers wrote his Aberdeen solicitor morosely, "I have tried several times to obtain a discharge from my creditors. I made as full and complete a surrender as was ever made by

Above left: Site of the Isabelle Tunnel. The tunnel mouth is currently concealed by trees at lower left of photo. (*Courtesy of Judy Wickwire*)

Above right: Hold-downs that may once have secured compressors or boilers near the mouth of the Isabelle Tunnel. (*Courtesy of Judy Wickwire*)

any man—handed to my trustee enough to pay every one." In a separate letter to attorneys representing creditors, he begged for the old debts to be forgiven, "on my family's account if not my own."[119]

As if that were not trouble enough, discontent over Chalmers's management was bubbling up again in London. Lord Poulett and two associates were criticizing his accounts, and Chalmers worried that a movement was afoot to get rid of both Syme and himself.[120] Syme seemed to be feeling the heat even more acutely than Chalmers; "[I] am truly sorry to find such a run against poor Syme," Chalmers confided to his son Willie. "I think Syme has been shamefully treated."[121]

Putting on his diplomatic cap, Chalmers responded in painstaking detail to a lengthy list of questions from the company's accountant. Among other things, he stressed that the company still owed him over $30,000, and that he had taken on personal debt of more than $3,000 to buy supplies for the Boarding House on the company's behalf.[122] Rather than apologize for his financial statements, he boldly demanded that from now on, his accountings be "audited and approved every month" so there would be no further question.[123]

Chalmers also penned an exhaustive letter to the Isabelle secretary, defending himself and his managership. His services were a bargain, he emphasized; by contrast, "Sutro has $1,000 per month [and] a fine mansion to live in, for caring for which Mrs. Sutro has $500 per month."[124] As for why Chalmers had recently declared his intention to become a U.S. citizen, this had been "very much against my inclination," he assured them, but was a step to ensure that he could hold patents, a privilege denied to a foreign corporation.[125]

Thanks to the tunnel's new ventilation, he was able to tout speedy progress of 300 feet per month.[126] By the end of March, the Isabelle Tunnel was 2,100 feet "in hard rock." Turning their own criticism back on his critics, Chalmers sniffed: "Our Consulting Engineer can now point out what can be done by a *makeshift* Manager and *inefficient, inexpensive* Machinery."[127]

He was hardly content with the equipment he had, however, and still angling for improvements. His men were currently using the popular Blake's rock crusher, for example, an invention that had been around for more than two decades. In early March, Chalmers wrote off to a foundry in Leeds, England, inquiring about a newer-style rock breaker he had just read about.[128]

Despite the recent influxes of cash, Chalmers had blown through them nearly as fast as they arrived. By early March 1880, he was pointedly advising London that he had only enough capital to continue operations through the end of the month. He also reported grimly that four more workmen had pulled up stakes and left due to the bad air before the ventilation work was completed. Making life even more difficult, there was 6 feet of snow now on the ground.[129]

As always, Chalmers wove in encouraging news to leaven the bad. He urged Syme to closely examine ore samples from a mine Chalmers wanted to buy near Monitor: "In one of the parcels of ore you will find marked a very rich piece of free gold."[130] He also had his eye on a gold mine in Tuolumne, California, which he hoped to pick up for his friends in London for $40,000 to $50,000.[131]

But work was not all that had been occupying him. Chalmers had spent some of the previous winter tinkering at his wood-turning hobby. Pleased with his

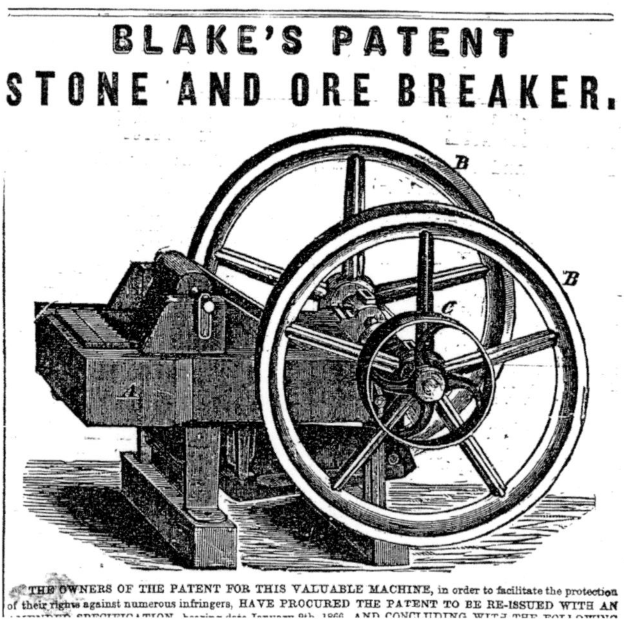

Advertisement for a Blake's rock crusher. This particular model was patented on 9 January 1866 and featured upright, convergent jaws with a revolving shaft and flywheel. By adjusting the distance between the jaws, the operator was able to "produce [rock] fragments of any desired size." (*Michigan State University Archives and Historical Collections, James Jones papers, collection c.103*)

success, he shipped off shirt studs made from local mountain mahogany and a small wooden hairpin box, "both of my handiwork," to Antoinette's friends in San Francisco, Mr. and Mrs. Kelly.[132]

Antoinette, now the unofficial mistress of the house at Silver Creek, was busy putting her own touches on the place. She had already picked out a new bedroom carpet on one of her forays to San Francisco.[133] In April, a two-page list of new cookware was ordered.[134] Next on her wish list was fresh wallpaper. Chalmers asked a San Francisco dealer to send samples of a "neat, cheerful, cheapish Bedroom paper, [with a] small pattern," cautiously insisting that it be "free of poisonous green."[135]

Another £1,000 remittance dribbled in on March 28, 1880.[136] But by late spring, Isabelle's finances were already getting tight again. Chalmers did his best to economize. Instead of buying a new blower, for example, he sent his bookkeeper to Virginia City to examine a used one. He also tried again to sell one or both of his hoisting rigs, though without success.[137]

As spring progressed, his letters show Chalmers getting short with merchants to whom he owed money. To one, he complained: "the last pipe [you sent me] was poor quality of iron, very brittle."[138] To another, he bristled: "You have charged too much for solder and $1 too much for the gauge and the hose is very dear."[139] He whined about a half-cent price variance to another merchant and complained more generally to L. Reynolds of San Francisco, "You charge like the mischief."[140, 141]

He also had to contend with products that had been damaged in shipping, or were poor quality in the first place. A carboy (jug) of muriatic acid arrived broken and empty, for example, while a shipment of car oil proved "worse than useless."[142, 143] When some 56 pounds of coffee disappeared in transit, Chalmers made sure to deduct twenty cents a pound from his freight bill.[144]

A late snow closed the roads and delayed arrival of much-needed supplies. By April, the "mules are getting pretty low down from scant rations," Chalmers chafed.[145] But every day the tunnel continued to lengthen, and by April 26, 1880, it stretched to 2,471 feet.[146]

Although the ventilation problem had ostensibly been solved, woes with the blasting powder continued. Bandmann's product was proving "utterly worthless," Chalmers complained, adding that "it has no power," and even with the improved ventilation system, fumes were occasionally still problematic.

"What is the matter with the last batch of No. 1 you sent me," Chalmers huffed to the Vulcan Company, describing his men as all "more or less stricken down with the gas." His foreman, he reported, suspected that "you have been keeping back glycerine and putting in acid."[147] Another expert believed that "frequent thawing and freezing had weakened the Powder."[148]

All of those difficulties wasted time—meaning slower progress in the tunnel. It did not help that the Isabelle was now into the "meanest kind of blasting rock." Once an array of sixteen to twenty-one 8-foot holes were drilled in the face, they now had to be loaded with about 100 cartridges—packed with eighty pounds of Giant Powder—to shatter the rock.[149] But at least his latest innovation was working out well: an "electric blasting" system using 8-foot fuses of gutta percha-coated copper wire, detonated by a Laflin & Rand Magneto No. 3 battery. This added to the cost, of course; they went through 1,800 of these fuses in a single month.[150] To deflect potential criticism of this expense back in London, Chalmers took the unusual step of having his foreman verify in writing that the rock was "harder than anything we ever met in Sutro."[151]

By May 3, with no remittance since late March, Chalmers's bank account was some $3,400 in the red.[152] Once again, he found himself in a predicament. "Funds are now exhausted," he wrote. "Am I to continue driving."[153] Syme was still trying hard to raise funds; a meager £500 floated in a few weeks later, followed by a token £100.[154] But on May 31, Chalmers chided: "No remittance to meet Payday tomorrow. This won't do."[155]

The lack of funds did not seem to stop Chalmers from continuing to investigate new equipment. He inquired about the cost of a "cupola" that would allow him to "cast my own Car wheels and other small castings," and waxed eloquent about the potential of a Diamond drill for core prospecting.[156] "Indeed a Tunnel such as ours should not be without one," he assured his friends in London. "Sutro has half a dozen."[157] He also inquired about the price of a mine locomotive to haul rock.[158] Nobody could accuse Chalmers of thinking small.

Although most of the old Exchequer creditors had by now been bought off, the Scossa brothers (local timber contractors) were still owed over $5,000 plus damages on their unfinished wood contract—and had secured a judgment for $1,800 of that.[159] The Scossa brothers were among the few creditors Chalmers seemed to genuinely fear; he would later describe them as "dangerous Italians" and "cut-throat looking churls."[160] The Scossas had been badly burned by the Exchequer default. Staunchly honorable men, they had been forced to go to work for other contractors to pay their own workmen. They were, quite understandably, still unhappy about it.

Whether the Scossas made an overt threat or not is unclear, but Chalmers warned London they might sabotage the company's property if left unpaid. "Arnot says they are revengeful and might harm the Mills," he wrote Syme.[161] "I don't think it

Adolph Sutro, *circa* 1880. (Matthew Brady photograph, Library of Congress)

prudent to run the risk of fire," he added later.[162] Perhaps more to the point, their $1,800 judgment included Chalmers personally, and "my sticks of furniture are in danger until he is settled with."[163]

Chalmers managed to compromise with the Scossa brothers in May, paying half their claim in cash ($2,500, in installments of $500 per month) and $638 in debentures for each brother in lieu of the balance. For the Scossas, this was still a losing proposition, though better than nothing.[164]

To Chalmers's relief, additional funds arrived in July and August, with $2,436 and $7,272 wired to his account with the Bank of California. It was enough to allow him to make a "clean sweep" of all the remaining indebtedness, except that owed to store-owner John Sauquet, "which is too large and too well secured" to be easily compromised.[165] Sauquet's bill would simply "have to be dealt with when Imperial is re-organized," he concluded.[166]

Still, on August 16, less than a week after the second remittance arrived, Chalmers was requesting that the company send a "larger remittance than usual" to enable him to lay in winter supplies. He had already jumped the gun on that, ordering a carload of powder he had no money to pay for; but 35,000 to 40,000 lb of powder would be needed to carry on the work through the winter, he emphasized. He had also ordered 1,000 feet of track iron and 1,000 lb of lime, and let it be known that he was dealing with a broken compressor.[167]

Nothing But the Best

By now, Chalmers had achieved truly astonishing progress in the tunnel. The Isabelle was nearly 4,000 feet into the mountain, more than halfway to its original goal, and crews were continuing to add roughly 300 feet per month.[168, 169]

Cautioning that "jumpers of timber are abroad," Chalmers patented 160 wooded acres to secure the property for Exchequer and I.X.L.[170] He also pursued the purchase of an "utterly worthless" 40-acre parcel for a very practical reason: to "prevent Saloon Squatters near my Tunnel." Enterprising liquor merchants, it seems, had set up shop where thirsty miners would not have far to walk for a drink. Chalmers was having none of it, of course. Not only did he try to buy the land, he briefly contemplated "making a town site of it," he wrote his lawyer in Sacramento, though he quickly crossed out the wild notion.[171]

In late summer, he ordered a new engine, likely to drive equipment at the tunnel. This gigantic piece of machinery clocked in at 10,936 lb—so heavy that Chalmers wrote the freight company to notify them in advance of the engine's incipient arrival, and warned the road-keeper to have all the bridges "secured and the road in good order" for this extra-heavy load.[172]

Again displaying his early adopter colors, Chalmers inquired that summer about the cost of that novel invention, the telephone—two phones, in fact. He hoped to install one connecting his tunnel with the engine room, and another linking the tunnel with his house, a mile and a quarter away.[173] These evidently were installed; a century later, the old Exchequer house was said to contain one of the very first telephones, connected to the mine—along with indoor plumbing and an early version of a water heater.[174] In August 1880, Chalmers also wrote the California

Electric Light Co. in San Francisco, eager for particulars about their new "Electric Motor and Electric Lights for Tunnel use."[175] He was still entertaining the purchase of a locomotive to haul debris from the tunnel.[176]

For his own table, Chalmers once again spared no expense. In addition to household essentials like starch and bluing, he ordered "one case of good Lobsters" and a "case of best Holland's Gin."[177] Since a good drink requires a good smoke to go with it, he also placed orders for a large sack of Olive Smoking Tobacco; three boxes (100 count each) of "as good cigars as you can send me at about $5 or $6 a box" (sent express); Lorillard's Drawing Room long-cut smoking tobacco; and 100 American Girls cigars at $6 for the box.[178]

Antoinette's youthful enthusiasm, however, was rubbing her employer the wrong way. She had gone to a Fourth of July Dance, "and hasn't got over it yet," Chalmers complained over a week later. "[I]t would almost look as if this climate no longer agreed with her," he fumed, threating to "send her off to San Francisco for good."[179] A subsequent letter suggested he was trying to dispatch her to St. Gertrude's Academy in Rio Vista, California, "to improve herself in penmanship, composition, grammar, style, music, etc. etc."—a gentle way of washing his hands of her.[180]

Other letters, however, show him still eager to please, buying Antoinette a fur-trimmed "sacque" (jacket)—and an expensive one, at that. In signature fashion, however, Chalmers soon found fault with the item, complaining to the vendor that the fur was coming off and it was not "worth more than $75" (roughly $2,000 today).[181] Meanwhile, Antoinette was indulging her own buying whims, purchasing a set of "walnut cane-bottomed chairs." Chalmers would add maroon Moroccan cushions to them several months later.[182] He also ordered a tuning key for Antoinette's new piano, inquiring whether the seller knew of a tuner in Carson City.[183]

This cane-bottom rocker on display at the Alpine County Museum came from Chalmers Mansion, and may have been among the "walnut cane-bottomed chairs" purchased by Antoinette Chalmers in 1880. (*Dustman photo*)

To a San Francisco doctor Chalmers was trying to persuade to relocate to the mining camp on a monthly retainer, Chalmers bragged that he "never require[d] a doctor," and thus had spent nothing on medical attendance for the dozen years he'd been in Alpine.[184] But perhaps thanks to all that fine food, he was growing stout. Chalmers wrote his tailor complaining that the drawers he had made were "too tight in the legs" by two sizes.[185]

Late August brought news of a most unexpected sort from Bavaria: daughter Laura was converting to Catholicism. Chalmers should, perhaps, have expected it, given that he had sent her to a church-related school, but he was horrified.

> My own dearest Laura,
>
> Nothing short of the announcement of your death could have distressed me so much as did your letter, received yesterday.
>
> It was only the other day that someone said to me, 'Are you not afraid your girls will turn Roman Catholics?' Little did I think when I replied, 'No, my eldest girl has more sense than that and she will keep the others right' that my confidence had been so misplaced, and that my false security was so soon to be disturbed. I have been fondly looking forward to the time when we should all be together again, and you the head of the house in your dear mother's place. I do so no longer, for no Roman Catholic shall ever rule in my home.
>
> I might have known what would be the result of sending you to such a school. These proseletizing heretics know well how to mystify and entangle with their Jesuitical sophistries even strong-minded men and women. When was it, dear Laura, that your father's wishes had no weight with you....
>
> Pause, I abjure you by the memory of your saintly mother, before ruining your own—all our happiness, by adopting a religion which your father has thrice solemnly sworn to abjure.... If, however, you care so little for your father or his happiness in his declining years as to be led by the sophistries of smart, designing men ... or the well-meant pleadings of misguided, church-enthralled nuns who are merely the creatures of those priests, I pray you do not attempt to beguile your sisters—whom it shall be my duty, the moment I can afford it, to remove from an atmosphere so pestilential.
>
> Your Aunt sowed the seeds by taking you against my wishes to High Church services, and she has paid the penalty. I pray God, however, she may be restored to her sane mind.
>
> Vexed beyond measure, I remain now, as ever, my dearest Laura
>
> Your affectionate father.[186]

Work on the Isabelle tunnel was continuing apace, and as the tunnel grew, so too did the need for materials. In late September, Chalmers ordered 3,600 additional feet of track iron, and put in an urgent request for borax and 15,000 shingles.[187, 188] But there was now no money left for much-needed powder—or payroll. Chalmers wrote tartly to the Isabelle secretary: "I cannot run the Tunnel in this way. Wednesday is pay day."[189]

In response, the company wired £1,500. Chalmers promptly asked for £500 more.[190] His September report noted caustically: "I have been obliged to overdraft at Bank of California, and anticipate trouble," chiding that remittances were needed on "the first, not the last, of each month."[191]

Once again, Chalmers added a touch of good news to his report—though it included plenty of bad. Using two drills, the men had bored 8,460 feet of 2-inch blasting holes and removed 2,798 tons of broken rock, "so tough that we had to charge and blast our centre cut holes twice." Despite the difficult rock, the tunnel had progressed another 328 feet.[192] They had finally reached the Balaclava Ledge, though they found it "broken up by later formations," a euphemism suggesting the much-anticipated ledge was barren. What they had struck was water. The workmen now had to fight water "literally gushing from face and sides" of the tunnel.[193]

Not long after Chalmers sent his disappointing report, he finally received permission to buy the Diamond Drill he'd been coveting.[194] But he, too, now received bad news mixed with the good. The London directors were again questioning whether their money was being well-spent. They had determined again to send one of their number out for a personal inspection of the work—a man named George D. Cardew and, sadly for Chalmers, one of his chief critics.

Bristling at this evidence that the directors were unhappy, Chalmers wrote asking pointedly "in what respect I have offended or failed in my duty to this Company or its Directors." Unbowed as ever but concerned with his "precarious tenure," Chalmers added his own demand to be confirmed as managing director.[195] It was a subtle way of reminding London that his services would be difficult to replace.

Cardew arrived at Silver Creek on November 7, 1880, accompanied by a second mining expert, William Ashburner. They would remain with Chalmers for five uncomfortable days.[196] Ashburner had been a member of the famous William Brewer geographic survey party that trekked California in 1863, and had studied mining in France.[197] Like Cardew, he would prove no friend to Chalmers.

After the pair left, Chalmers responded to follow-up questions from Ashburner, giving particulars on his disastrous mill runs of 1876–77. The best result from

William Ashburner received his mining education in France. Around the time he visited Chalmers, he also prepared an 1879 report on the famous Empire Gold Mine in Amador County, California (published in 1880). He would later become an honorary professor of mining at U.C. Berkeley, and served as a university Regent. He passed away in April 1887, less than a decade after visiting Chalmers. (*Illustrated Fraternal Directory, Including Educational Institutions of the Pacific Coast (San Francisco 1889) (WikimediaCommons)*)

the battery samples, he was forced to admit, had been $25 per ton.[198] He also was forced to defend some of the figures he'd sent to London, noting that air compressors had cost $384.24; $700 was spent on tunnel buildings; and some $1,084.24 of Isabelle funds had been diverted ("lent") to Exchequer.[199]

To nobody's surprise, least of all Chalmers's, Ashburner's report was "damning."[200] But Chalmers still had powerful friends in London, and he would get the last laugh in the end. Nearly a year later, Cardew was pitched off the Isabelle board of directors, and he and associates were forced to surrender 40,000 Isabelle shares at a "nominal price."[201]

Wedding Bells

Antoinette had been suffering for months from what Chalmers described as "fainting spells that alarm everybody."[202] Her condition worsened, and that September, Chalmers wrote to Mrs. Kelly that Antoinette "has been at death's door with these fainting attacks which have kept her very weak. For several nights, we have taken turns to sit up with her all night."[203] By October 1880, she was suffering "convulsive fits," one of which lasted three days.[204]

He penned a lengthy letter to a doctor describing her symptoms, "[l]est she should omit to impart to you all that ailed her":

> I think it was about the 17th of last month [September] that, being in her usual health, sitting chatting with some friends from Columbia who were paying her a visit, she became suddenly faint and called for water, which she got, but immediately fell senseless to the floor. Hearing the noise I hastened to see what it was, carried her to bed, had her clothes unloosed, and, by dint of cold sprinkling and vinegar and water applications to her head (of pain in which she always complains), succeeded in restoring consciousness. But from that [time] on, she was out of one fit, into another.
>
> Commencing with a fixed look of the eye, a slow but strong and determined stretching of the arms, convulsive twitching of the fingers, gradually reaching toward the mouth, biting the hand savagely (of course [we] prevented, by giving [her] a wet towel to bite), raising the body from the bed as rigid as if marble, with a strength which would surprise you—trying to bite, all the time features distorted, teeth clenched until, exhausted, she would sink back on her pillow, teeth set, eyes fixed and frothing at the mouth and apparently dead, though [her] body all warm and circulation pretty regular.
>
> Bathing the head with vinegar and water, applying ammonia to the nostrils would eventually bring her round, when she would say, crying—"Oh, I have been so sick; how my head is like to split; I know I am going to die!" [and] bid us all good-bye, ask the Bible to be read to her, leave Henry to my care, until by coaxing and fondling I would get her to feel that she was not so near death. Often when they came back they are preceded by a cynical smile—[she] says she sees her friends on the other side of [the] water and, ushering us out of the room, then the convulsive fits [begin]—slow, rigid stretching of the limbs etc. etc.
>
> These continued for three nights—worst at nights. She would get up and go about during the day and be as cheery as ever, except the first day after the first night's attack, she lay in bed till dinner time, after eating which she had another attack. And I

> have observed that it is generally after eating (she has a ravenous appetite sometimes) that they attack her; once after drinking a glass of Tenant's Porter, once after [eating] oysters, often after dancing or great and unusual exertion.
>
> On asking her how she feels when they come on, she says I have no warning; everything gets black and dark to me; [I] have a pain in my stomach which gradually rises until it gets in the region of my heart, can't get breath, can neither see nor hear, head painful and then all is chaos, till she gets over the fit....
>
> Now, Doctor, unless you can cure her of these attacks, I do not think it right she should return here, where there is no doctor, and where she might die on my hands at any time. I wish, therefore, you would make yourself sure of what is the matter, and write me candidly. Of course I will do my best to provide for her in San Francisco, but the apprehension of these attacks coming on her here at any time makes her very uncomfortable and childish——she is positively in the way with her 'Oh, but she'll never get over this' etc. etc., crying, and making a d[amne]d fool of herself generally, injuring in place of aiding, frightening in place of assuring. Of course the attacks leave her weak, but she soon rallies, and is as cheerful as ever....
>
> You can say I wrote to you to advise her and see that she obeyed the doctor's orders.[205]

Chalmers had seemed content to remain a bachelor for many years, but the next month, he suddenly changed his mind and decided to marry Antoinette. It is hard to know whether this was prompted by sympathy for her medical condition or if something else led to his sudden about-face. Later letters suggest he was confident his finances were on an up-swing, so it is possible he felt able to give her more security. Marriage would also put a stop to her frequent sojourns to San Francisco, leaving him worrying each time about whether she would come home again. Or perhaps he simply decided it was time to silence the local gossips.

Like pretty much everything else Chalmers did, once he had made up his mind, he did it up right. He and Antoinette tied the knot on November 30, 1880, at the magnificent Grace Church Cathedral in San Francisco.

Antoinette must have felt like she was on top of the world, her future secured at last. She wore a beautiful new dress for the occasion, a fashionable dusty-pink known as "ashes of roses" that complemented her dark curls. Their honeymoon, short though it would be, was spent in sparkling San Francisco.

Looking back, Chalmers would later sigh, "When I married, everything looked as if I were to make a brilliant success."[206] But on December 14, just two weeks after the wedding, everything changed. With the Isabelle tunnel now over three-quarters of a mile long, Chalmers received the horrifying order from London to stop work.[207] The tunnel had been pushed 4,429 feet in just eleven months and ten days.[208]

It was the worst of all possible timing, in more ways than one. Not only was Chalmers newly married, but the tunnel was almost within sight of its goal, with only another 1,000 feet or so to reach its hopefully-rich lodes. Taxes were again due, and Chalmers was again without funds to pay them—or to pay his men for November.[209] His new wife, Antoinette, must have been worried sick.

Chalmers dispatched a frantic telegram to London: "Miners mutinous. Cable immediately six thousand"—the amount needed for forty-five days' wages.[210]

Antoinette Laughton in her wedding dress, *circa* November 1880. The dress was the color of "ashes of roses." Antoinette was about twenty-seven at the time; Lewis Chalmers was fifty-five. (*Alpine County Historical Society*)

He was not kidding about the miners' "mutinous" sentiments. Three days later, Chalmers sent an even more urgent letter to the Isabelle's secretary:

> [T]he men are on a drunk and threatening attachments, personal violence, and all manner of things, unless paid, and no money to pay with.... [I]t got rumoured about that I was about to go away, and the sleigh carrying the mail was stopped to take me out by force.... There will be trouble here unless you pay up—of more kinds than one.[211]

The newlyweds passed an anxious Christmas together. Sure enough, the first wave of trouble was not far away. On December 28, 1880, a fresh lawsuit was filed. The Vulcan Company, wiser from sad past experience, had decided to wait no longer for the $8,000 Isabelle owed for powder.[212]

10

A Leap of Faith with Morning Star (1881)

Looking back, Chalmers would blame both a "villain named Cardew" and "dissension among the Shareholders at home" for halting work at the Isabelle.[1] Stopping work was both unfair and illogical, in his view. After all, the tunnel had already been pushed over 4,400 feet into the mountain, with only another 1,000 or so feet yet to go. Cardew and Ashburner, he felt, had been biased against it from the beginning.

But for now, at least, it was clear that the once-promising Isabelle Tunnel was a dead horse—going nowhere. Chalmers might have been forgiven if he had simply tendered his resignation and departed for home. That thought must certainly have crossed his mind. But Chalmers had invested too many years in his Alpine mining dreams to throw in the towel.

True to form, he battled to keep his post. Also true to form, he concocted a brand new plan to rescue the Isabelle's flailing finances—by turning his attention to a new set of mines. He had been keeping an eye for years on several mining claims near Monitor. These, unlike the untested lodes he had previously pursued, had been actively worked in the past and (at least from time to time) had produced profitable ore. That made them both more promising as a source of badly needed cash and more likely to draw fresh interest from London investors.

In February 1881, Chalmers outlined for Syme the bare bones of a plan he hoped would spell salvation: buy the Colorado No. 2 (to be renamed the 'Pluto') for Isabelle Company; the Morning Star (or 'Stella') for I.X.L.; and the Advance for the Exchequer. Ore from these three well-known mines, he contended, would enable him to ship bullion almost immediately. The excitement generated by displaying actual bullion would boost the price of the new venture's shares from one pound to five "in thirty days"—which in turn offered the "[s]implest way to obtain cash for [Isabelle] tunnel."[2]

Actually putting that plan into practice, however, required months of planning and wheedling. Naturally, more money would be required—some £8,000, according to Chalmers's estimate. In early April, Chalmers again pressed the Exchequer board to pursue the "immense advantages which must accrue to them if they adopt my suggestion"—advantages that he predicted would quickly include dividends.[3]

Illustration from *The Family Cyclopoedia of Useful Knowledge.* (*F.M. Lupton Co. 1885*)

While the powers that be back in London were mulling over this new option, Chalmers continued to plug away at day-to-day affairs in Alpine. He ensured that the I.X.L. and Exchequer taxes were paid, using money he'd received in December.[4] He also embarked on a damage control campaign to respond to allegations in Cardew's report. This included collecting affidavits confirming that his secretary, George Jones, was not, in fact, a drunk ("I myself will swear that if there is a Sober man in California he is the man"). Oozing wounded dignity, Chalmers also denied the claim that he had been making "underhand profits" at the company's expense by charging boarding fees to his miners. These amounted to just a dollar a day per miner, Chalmers pointed out, the same cost charged by other boarding house keepers, and "[i]f I make anything I do not suppose it is any of Mr. Cardew's business; if I lose I don't suppose Mr. Cardew would pay the deficit."[5]

Chalmers added his own complaint about Cardew's unhelpful behavior while in Alpine, citing the man's "unaccountable statement to the Sheriff of Alpine that I had received money to pay [old claims] in full," which had "prejudiced my negotiations for the two or three outstanding claims mentioned ... and is also calculated to injure me personally."[6]

There was no one likely happier than Chalmers when the tide turned in late January and Cardew was pitched out as a director, "thanks to the manly stand taken by our noble Chairman."[7] That did not end the unhappy matter entirely, however, as litigation soon followed. Cardew sought to recover both the cost of his trip to Alpine and the money he had paid Ashburner for his "damning report on our Alpine mines."[8] The Isabelle Company, for its part, began totaling up the damages Cardew had caused by halting work on the tunnel—some $18,000, as they figured it.[9] It would be months before the dispute was finally resolved by a settlement agreement.

Although Chalmers was stuck twiddling his thumbs as the Isabelle sat idle, there was one bright spot on his horizon: by April, it was clear that Nettie was expecting. This was to be her second child, and Chalmers's tenth. To please her, Chalmers ordered flower and vegetable seeds. Radishes and crookneck squash, morning glories and hollyhocks would soon grace her garden.[10] Once again, he sought household help to ease Nettie's workload, specifying "a strong Scotch or German steady woman, unmarried or widow, [but] no Irish or American."[11] In June, he also ordered a length of decorative black walnut "for the top finish to a Bath" inside the house.[12]

He was not neglecting his own comfort, either, despite the doldrums enveloping his mining projects. In March, he had written to stable owners Keyser & Elrod in Carson City, asking if they had a "light spring wagon for two horses to sell cheap."[13] He apparently found a buggy to suit him, as by May he was sending for a buggy harness decked out with silver-plated buckles.[14] So much for his earlier claim of eschewing tobacco; he also sent away for 100 more cigars and a large sack of smoking tobacco.[15]

That summer, he ordered Nettie some nightgowns, ladies' house slippers, a length of white cashmere, and a paper of diaper pins.[16] Larger purchases soon followed, including a new bed and washstand, bureau, mirror and carpet.[17]

As the baby's arrival drew closer, Chalmers began making detailed preparations. Once again, cost seemed no object. He ordered an expensive bassinette crib "with Canopy" for $22 (over $600 in today's dollars) from retailer J. Samuels in San Francisco.[18] Having already lost one wife to complications of childbirth, Chalmers was not about to lose another if he could help it. In July, he ferried six-month-pregnant Nettie all the way to San Francisco to see a doctor.[19]

He wrote to a well-known Carson Valley midwife asking her terms and whether "you require the assistance of a medical man."[20] He penned a separate letter to Dr. Luce of Genoa, asking his recommendation for a "skilful female *accoucheur*," and—with the baby expected in September—the doctor's charge for arriving at Silver Creek on August 25 "and remaining until all is well."[21]

In part, that was simply being practical; Silver Creek was, after all, a long way from medical help if things with the birth went awry, but such caution also shows Chalmers's level of concern. Many a wife in Silver Mountain or Markleeville delivered their babies with only a comforting neighbor's hand to hold. Antoinette, however, would have the very best maternity care money could provide.

Snapping Up the Morning Star

Frustrated that the London directors were continuing to dither over his new proposal, Chalmers took a leap of faith in July and bought the Morning Star without their approval. This prominent mine near Monitor and had recently been relocated by Sheriff Scott and friends, and had successfully yielded both silver and copper in the past. To Chalmers, the price was simply too good to pass up: just $4,000 down, with another $3,000 due in a year.

Chalmers jumped at the chance, taking title in Syme's name. He had sold some of his own investments, he said, and "borrowed from Tom, Dick and Harry" to put the deal together.[22] At first, he advised Syme that he hoped the two of them might be able to float this new venture themselves. "If I can do this without aid from your side I agree with you that we should keep this to ourselves and be independent of Companies," he wrote, "as [the broker] would get all the cream, our stock being postponed."[23] This attempt at a quiet side deal did not pan out, and the new mine became an Isabelle property—despite being acquired without the board's permission.

> I took it upon us to buy the Morning Star with money in my hands belonging to this company [the Isabelle], and got wigged for my pains, which don't matter, as I shall come out all right. In three weeks from now I shall run ten stamps (Exchequer [Mill]) on this ore and have a profit of $100 per day.[24]

The directors, Chalmers was confident, would eventually be persuaded that he had done the right thing. "[B]ut say nothing to them officially until you secure [director] Leeming's interest," he admonished Syme.[25] He would later defend his rash step by explaining that he felt "at this juncture, something had to be done to aid the [Isabelle] Company," and cavalierly brushed off London's demand for a full report: "I do not waste time on a Report at present—Bullion will do instead."[26]

The Morning Star had been worked substantially in the past, and its tunnel once stretched over 1,600 feet long.[27] But the mine had not been operated in over a decade, and the repairs needed were substantial. Chalmers described the tunnel itself as "entirely caved in," and large sections needed retimbering. Some 100 feet of water filled the shaft and would need to be pumped out. Chalmers also planned to move the old Exchequer hoisting works down and reinstall them at the Morning Star, another big project.[28]

Pleased to finally have something productive to do, Chalmers launched a flurry of activity at the Morning Star. He set crews to work immediately clearing debris from the tunnel and improving the road leading to the mine. By mid-August, he was able to boast that 3 miles of new road had been constructed and 6 miles of existing road repaired, though "I still have some bridges to strengthen and cover to carry heavy loads."[29] When Syme complained about the money spent on road work, Chalmers retorted: "Don't be such a damned fool. Without a road, how was I to haul the ore here."[30]

The ore from the Morning Star (which Chalmers renamed the "Stella") was extremely high in sulphur, and the Exchequer Mill would again require modifications to effectively work it. Chalmers promptly set about adding a series of vats, tanks and agitators to treat the Stella ore by what he called the "Hunt, Douglas and Stewart process."[31] He was confident that this beneficiation process would work, assuring Syme: "Have no fear about treating the ore. We have tried it on a small scale and got excellent results."

Although he was "very busy getting the Mill ready," Chalmers fretted in mid-September that he had "nothing with which to do it."[32] He begged Syme to see if he could not round up a few new investors for a one-eighth interest. "Send me $11,000 to aid me in Stella," he urged. "Depend upon it, this is going to be a 'big thing' if I can only manage to get started."[33] Even the man installing the new equipment "says I have got my fortune in my hands," Chalmers assured Syme, adding that he firmly expected "to take out $300 at least a day, net" once the operation was up and running.[34]

Happy News and High Expectations

The big news that September had nothing to do with mines or silver. On September 20, 1881, Antoinette gave birth to a 9-lb baby boy in the house at the Exchequer Mill. Chalmers was every bit the proud papa, writing the *San Francisco Chronicle* to insert a birth announcement, and requesting six extra copies of that newspaper.[35] A few days after the baby's birth, he ordered a hair mattress, feather bolster, pillow, blankets, and spreads for the "little crib," noting happily: "Mrs. Chalmers last Tuesday gave me a fine little fellow to fill it."[36]

Even the excitement of his new son's arrival barely slowed Chalmers's push to launch the Stella, however. Just three days after this blessed event he was shipping off a box of ore to Syme, taken "from head of winze."[37] Unfortunately, when assayed in London, those samples proved horribly disappointing: just 15 cents per ton in gold, $9.75 in silver, and 15 percent copper. "There must be something utterly wrong about this, as I will soon show," Chalmers chafed. "But even 15% copper alone will pay handsomely if we had plenty of stamps, vats, etc."[38]

Chalmers countered with his own assays of ore from the same location, treated in San Francisco by the Hunt, Douglas and Stewart process. These painted a wildly different picture, showing $92.88/ton silver and $20.67 in gold, with 40 percent copper.[39] To follow up on this far more encouraging data, he appended a lengthy financial summary—essentially his "wish list" for further improvements. After estimating that the cost of mining and milling would not exceed $15 per ton "if these operations are conducted on a proper scale," he sketched out cost estimates for hoisting and pumping

This wicker baby crib once held Lewis and Antoinette Chalmers's first child, also named Lewis, born 20 September 1881. It is currently on display at the Alpine County Museum. (*Author photo*)

equipment ($40k); sinking 600 additional feet ($24k); drifting work ($10k); erecting a fifty-stamp mill ($60k); installing a wire tramway ($10k); and building boarding houses at the mine and mill ($5k).[40] "When hoisting works are erected, 50 tons a day can be extracted to supply a fifty-stamp mill," he confidently predicted.[41]

For now, with no mill yet in place at the Morning Star/Stella, 40 tons of ore had already been hauled to the Exchequer Mill, worth between $25 and $200 per ton, by Chalmers's estimates. With ten men working the mine, he was confident they could add another 10 tons to that total per day, "which is about all I can manage at this mill."[42] Even low-grade ore could be profitably worked, he assured Syme, so long as there was enough of it. "When you get into your head that nature seldom gives even $40 ore in any quantity and that the most productive and best paying Mines are those that have large bodies of low grade [ore]," he lectured, "you will think the Stella ore no subject for cynical remarks."[43]

Work continued apace through the fall to retrofit the old Exchequer Mill to work the recalcitrant Morning Star ore. Twenty-one new wooden vats arrived in mid-October and were quickly assembled, though not without some difficulty. Chalmers complained, "[the seller] gave us a very poor job. I have had to remake the tubs to get good joints."[44] He also set men to rebuilding his old assay office which was, as he put it, "falling down."[45]

Things were progressing on the London front as well. Syme sent along a draft "Prospectus of the Stella G.S.M. Co. Ltd." for Chalmers's comments and corrections, confirming his partners were trying to float this new investment.[46] Nevertheless, the London investment news had been critical and Syme was growing nervous about the success of the venture. Chalmers retorted tartly, "*Mining Press* I can't muzzle. The only reply to all your facetious criticisms I can make (and I will not stoop to reply otherwise) will be bullion, which I am trying to ship without your aid."[47]

By mid-November, Chalmers was preparing to start up his new reduction works in earnest. In its rebuilt configuration, the furnace now measured 64 feet long inside, with two hearths for roasting the ore.[48] To assist in getting out ore, he hired a new foreman and "four good [Cornish] miners," and for the mill he requested that feeders be sent from Carson, to "be here in about ten days."[49]

Winter weather had set in, bringing its usual difficulties. "My fingers are so cold I can scarcely hold my pen," Chalmers complained in a letter to his son, Willie, on November 21. It was not just Chalmers who was suffering. His mine engineer "got his feet frostbitten" and was laid up for a week. Two of the miners also decided to quit, possibly from the cold.[50]

The new baby, however, was "gaining in weight, and losing in temper," as Chalmers joked. "At night, he frequently keeps us all dancing to his music. I have made him a cradle and Henry [Antoinette's older son] is quite an adept at rocking it."[51] They had named the child Lewis, after his father; "[h]is mother would have it so," Chalmers insisted.[52] In truth, however, he seemed overly fond of naming offspring after himself; this would be the *fourth* child bearing some variant of "Lewis" in their name.[53]

By late November, the final pieces of his new project were falling into place and the renovated mill was almost ready. Chalmers had hired three additional ore teams and was eager to begin working Morning Star ore.[54] "Give me stamps enough and a mill and hoisting works and I will give you proportionate profits.

With forty stamps and everything arranged for the Hunt, Douglas and Stewart Process process I can give you from Stella alone over $1,200 to $1,600 profit per day. No doubt about this, Henry Syme," he wrote confidently.[55]

He sent off a fresh request for laborers and battery feeders, specifying that he wanted "hardworking and sober" men for $3 per day with twelve-hour shifts, "out of which comes one dollar for Board. They must bring their blankets," he added.[56]

The Exchequer mill finally was able to begin crushing Morning Star ore about December 15, 1881.[57] After "some trouble at the start with the furnace," he told a local merchant the following week "everything running nicely now ... all goes OK at last."[58] "[A]t long last I have got to running," he similarly confirmed to Syme, his optimism palpable. "It has been a hard struggle, but it is now over, and from now on, I hope and trust, all will be *couleur de rose*."[59]

The Isabelle Tunnel remained at a stand-still "in consequence of litigation at home," as Chalmers diplomatically explained it. More accurately, they were still out of funds. He continued to assure creditors he hoped to be able to pay them "before New Year's Day," but was forced to admit he remained "impecunious" at present.[60]

Even so, Chalmers had been quietly making inquiries about yet another Monitor mine, the Colorado No. 2—a portion of the famous Tarshish mine.[61] None other than capitalist George Hearst had been "after" it, as well.[62] But Chalmers's sometime-friend Arnot was a major shareholder and was working behind the scenes to try to craft a deal. "Arnot controls a great part of this ($30,000) and has been keeping them quiet, hoping to sell to us," Chalmers had assured Syme back in July.[63]

Chalmers finally secured a written offer that October to sell the Colorado No. 2 for $60,000, with a $5,000 down-payment due by December 31, 1881, and the balance in one year. "Wire if you can place it in time," Chalmers pressed Syme.[64] "Do this," he urged Syme in a letter two days later. "The mine will pay from the word go ... if you extend your batteries (another twenty stamps) and add four more concentrators."[65]

Chalmers managed to send home a favorable report on the Colorado No. 2 in mid-December, written by metallurgist O. Hoffman of the firm Kustel & Hoffman. Kustel had become famous in Comstock mining circles for his 1863 treatise on methods of extracting gold and silver, and the pair had invented a hyposulphite leaching process that had been successfully used at the Advance Mill in Monitor.[66] More recently, Hoffman had become manager of the Silver King mine in Arizona, and was, as Chalmers put it, "a very talented, honorable man."[67]

Hoffman's stamp of approval was a strong recommendation for the mine. Even stronger was the Colorado's own track record; as recently as February 1880, they had been "shipping Bullion pretty regularly."[68] Little wonder, then, that Chalmers believed this was another deal too good to pass up.

Dancing the New Year In

With Christmas coming on, Chalmers returned his attentions to domestic affairs. The baby was growing rapidly, he reported with satisfaction, though "Nettie is not strong."[69] Letters from home had brought the sad but not unexpected news that sister-in-law Eliza had passed away in November in an insane asylum in Bavaria.[70] Meanwhile, he had discovered to his disgust that son Johnny had been sent to a Roman Catholic seminary in Pennsylvania to become a priest.[71]

Chalmers implored Syme to send his son Alick out to join him, stressing that his surveying and drafting talents could be put to good use. "I want a man who is a good surveyor to reside at the Stella," he urged Syme. "Send him [here] at $100 per month and I will make him very useful. [He can draw] plans of mine and mill, etc. etc. etc."[72]

Chalmers was apparently in no position to help Alick with traveling funds, though, as once again, money was tight. In a letter to a former employee now at Virginia City, he complained: "At no time since my arrival here 13 years ago have I been so poor as I am to-day." The Exchequer still owed him $28,000, "which I fear I shall never see." Yet with characteristic optimism he added: "My friends have bought the Morning Star Mine, and we have just commenced crushing."[73]

Giving the lie to his supposed impecuniosity, however, Chalmers decided to throw a lavish New Year's Eve dinner dance, complete with festive Christmas tree. He invited a veritable "who's who" of Silver Mountain and Markleeville notables, including the Arnots, Sheriff McBeth and his wife, and the Monroes of Markleeville, urging them to share the word with "as many more as you choose." Also invited to attend were all his miners and their families. "Ask the boys to bring their music with them, and their girls," he instructed in one letter, "[and] Jarvis his violin." To ensure that his miners would have female partners to dance with, he urged two local widows to "[c]ome along and bring any nice girls with you you can pick up."[74]

To decorate the tree, Chalmers wrote away for "a box of candles, dolls, etc. etc. or variety of such things as you may think best, flags, mottoes, etc. etc."[75] He planned to play Santa himself, adding an order for a "Santa Claus mask or false face with long beard."[76]

He would later explain the event as an effort at employee relations, claiming he had planned it "to keep my men sweet, so gave them a dance at New Year." But no matter how good his intentions, Chalmers always found critics; the expensive festivity prompted "a flaming account" in the *Monitor Argus*.[77]

11

The Planks Are Getting Very Few (1882–1883)

Chalmers's old, familiar struggles would continue to dog him into 1882, but by January, the revamped Exchequer Mill was finally up and running. Initial battery samples of Morning Star (Stella) ore were impressive, averaging $62.25 to the ton, including $40 in silver and $6 in gold.[1]

In its current configuration, the mill could handle 5 tons of ore a day, but Chalmers confidently predicted that the Morning Star was capable of supplying 50 tons a day. Figuring an average yield of $60 per ton, that was a tantalizing prospect. Chalmers argued passionately that additional funds should be sent to immediately enlarge the mill—and thereby its profits.

"All we want is milling capacity," he pressed Syme on January 25, 1882, "for we have a mine ... which, were it on the Comstock, I could sell for $150,000 just as she stands."[2] He made the same pitch to the company's directors, projecting a "clear profit" of $200 per day if only they would furnish another $5,000 to double the mill's working capacity to 10 tons. "Let me conclude by assuring you that you have the best mine in California," he added smoothly, "but you must work it on a more extensive scale to reap the benefit."[3]

The company had only recently funded major changes to the Exchequer Mill, following Chalmers's earlier recommendation to adapt it to the O'Hara method. So the total cost of the mill already topped $45,000.[4] The directors were understandably dubious about spending even more. It was the unusually high copper content of Morning Star ore that had demanded those modifications, Chalmers argued, firing off a dissertation on metallurgy in answer to his critics:

> Base metal ores must be chlorodized before they can be leached successfully or amalgamated, *i.e.*, not only must all the sulphurets be deprived of their sulphur by an oxidising roasting, but they must be converted into chlorides by the addition of salt (chloride of sodium) after the oxidising roasting is completed. The O'Hara [furnace] does this perfectly on such ores as Exchequer and I.X.L. ... but when your ore contains so much sulphur as does Stella, you can't [with] any mechanical furnace keep it in as long as you wish to, to get all the sulphur out ... [W]ith a mechanical furnace, where taking in ores containing so much sulphur, you have to run your ploughs slow—our

Illustration of miners at work, from *The Family Cyclopoedia of Useful Knowledge.* (F.M. Lupton Co. 1885)

> chams [*sic.*] are always heating and stretching, ploughs wearing out, entailing many stoppages to repair, so that if I had a mill to build for Stella ore I should put in the old reverberatory with hand labor.... Again I repeat the O'Hara is perfect for the Exchequer [or] any ore containing the same amount of sulphurets. Again I repeat that the Stewart process is perfect for Stella ore. All we want is Capacity.[5]

But merely doubling the Exchequer's working capacity was not all that Chalmers had in mind. Once that initial step was accomplished, he explained, he hoped to add agitators and reverberatories and then boost its capacity even further, from 10 to 18 tons a day. Similar modifications to the nearby I.X.L. Mill would allow it to process an additional 20 tons daily, bringing the milling capacity to a whopping 38 tons of ore every day. His plan to move the old Exchequer hoisting works to the Stella would boost ore output. And with all that in place, Chalmers boasted: "if I don't make Alpine boom and put [the company] *in excelsis*, my name is not Lewis Chalmers. There is [*sic.*] no ifs or uncertainty about it. You have a grand Mine."[6]

Once again, the company cautiously decided to send out a mining expert to verify Chalmers's claims before advancing further funds. Rather than dispatch someone from London this time, they hired Captain Benalloch of Grass Valley to assess the operation. Additional financing would follow, they assured Chalmers, if Benalloch's report proved "satisfactory."[7]

Benalloch arrived in March 1882, and at least on the surface, he and Chalmers got along famously. Benalloch even hinted he might offer Chalmers the position of managing director for a gold mine he owned in Grass Valley. All signs seemed to suggest that a happy report would be forthcoming.

Benalloch "will tell you he never saw a better piece of work than the [Morning Star] Tunnel, or a finer mill than the I.X.L.," Chalmers reported confidently in a letter to son Willie in early March, declaring Benalloch a "good, honest fellow and no doubt a good miner."[8]

To Chalmers's chagrin, however, Benalloch's report delivered a less-than-ringing endorsement of the Morning Star, refusing to confirm Chalmers's assessment that the ore body was virtually unlimited in size. Piqued, Chalmers visited Benalloch in Grass Valley the following month to confront him in person about his negative report, though Benalloch protested he was merely trying to be cautious.[9]

Choosing Benalloch as the investigator had been "ruinous folly," Syme agreed, but the report, flawed though it was in Chalmers's estimation, effectively put an end to any grand plans to expand the mill—at least for now.[10]

With his finances pinched again, Chalmers faced the all-too-familiar difficulty of keeping workers on the job. Six of his mill men agitated to be paid in January—"Comstock union men," he described them in disgust. He promptly let all six go, determined to find replacements who would be "sober, competent, reliable" and thoroughly non-union men.[11] He unabashedly advised his new hires they would need to wait sixty days for their first month's pay.[12]

But labor troubles erupted anew the following month. "I had a rebellion among my miners yesterday and all my men left, and [it] took every cent I had to pay them," Chalmers groused to a creditor in mid-February.[13] Even his stable hand, tired of irregular pay, departed for greener pastures. In a telling letter seeking a replacement, Chalmers outlined the job and his expectations:

> Can you send me an A-1 (sober) Scotch or English groom who can drive, milk two cows, and bring in wood and make himself generally useful. He must be a thorough groom and up to keeping his horses and carriages and harness in apple-pie order. Wages $60 a month and he boards himself at Company's boarding house at $1 a day. To a first class man not afraid of work I would give $70. In summer he would put his spare time in the garden.[14]

Three days later, he added a short postcript:

> In writing you last mail about a groom I omitted to mention that I want a light weight [man], not over 150 or 160 [lb]—the lighter the better, in our roads. It is immaterial to me whether the man be married or not, if he leave his wife behind, but I prefer a single man, because it is not always convenient for me to pay on the first day of every month, and married men need their money.[15]

More distressing even than the difficulties with his employees, Chalmers received troubling news about his family. In April, the convent in Bavaria wrote to inform him that his eldest daughter, Laura, had passed away and younger daughter, Jessie, was "very sick"—news that came "like a thunder-clap."[16] "If I can't manage to get the other two girls [Jessie and Elleny] away, they will go next," he fretted to Syme. "I think the climate is damp and inimical to weak lungs, which I fear they inherit from their mother."[17] Brother John, too, who had previously shared Chalmers's home in Alpine, had "gone to his long home" a year or two earlier.[18]

Chalmers wrote anxiously to his two remaining daughters, imploring his "dear, dear girls" to "take care of your health, don't expose yourselves to damp—drink milk with a deal of sugar in it, and live in hope that this Summer I may have you here.... Meantime write me every week."[19]

Son Johnny had tired of his priesthood studies in Pennsylvania and begged his father to bring him west. Chalmers replied firmly, "When you give up all idea of being a Roman Catholic, write to me, and if I can find the means to bring you here I will do so."[20]

Chalmers was feeling his own disappointments keenly after fourteen long years of labor in Alpine. In a sentimental letter to a childhood friend from Scotland, he lamented that his efforts thus far had not been met "with the success which I had expected," and acknowledged he was "growing grey." He reminisced about their youthful escapades together in Fraserburgh, "knock[ing] the watchman's hat over his eyes on Union Bridge, and ... taking off the old couples' shutters, etc."[21]

Sharps and a Shoestring

Chalmers had managed to produce several small ingots from Morning Star ore in his first January mill run, but he had not yet perfected the processing technique. Those first silver bars were only "base bullion, ranging from .150 to .450 fine," although Chalmers expressed himself hopeful that he would soon manage to produce a higher grade.[22]

> But you must have patience until my calcining pits are going. Having storm[s] to contend with, and ground (when it don't storm) hard frozen, I have not made the progress with them that I hoped.... As, however, success is sure, I can stand it and there will be no collapse as you lugubriously predict. Help me a little if you can in the meantime.[23]

Later that same January, he had shipped $1,600 worth of bullion, by his estimate, to the bank in San Francisco. It was not a huge amount, but it would help defray expenses, and Chalmers hoped to be able to "continue at the rate of $5,000 a month."[24] Cement copper was also dispatched to New York in January, to bring in extra money.[25]

But even these apparent successes quickly morphed into disappointments. By late March, Chalmers had shipped what he estimated would be $5,255 in silver and copper, from which he expected to net over $4,000. Instead, he discovered to his horror that the "San Francisco sharps" demanded a steep 17.5 percent discount due the poor quality of the bullion. It was still, as he explained to his friend Wheeler, "too low grade to go to the Mint. But this I will overcome," he swore.[26]

With only a shoestring budget, he pressed ahead as best he could to finish upgrading the mill to the more copper-friendly Stewart process and expand the mill's capacity to a hoped-for ten to twenty tons per day. Among other alterations, he added four, triple-deck hand reverberatories and enlarged the brick roasting oven.[27]

But every penny spent on the mill meant less in his pocket to pay for other things. Strapped for cash, Chalmers even sent Nettie's physician a post-dated check. "I have been obliged to date the cheque next month," he wrote without apology to Dr. Luce, "but Livingston [a prominent Genoa businessman] will no doubt discount it for you."[28]

Syme managed to send a tiny but much-needed infusion of $482 in late March, which Chalmers used to pay his most pressing creditors. He responded tersely:

"It's simply hell."[29] The following month only deepened his financial troubles. "I am in hourly danger of an attachment," he confided to Syme in late April, "and once commenced, my men will strike and commence hostile proceedings.... I will do my best. But it's pretty hard times."[30]

Reluctantly, Chalmers departed for London again in late May to seek funding. He had been forced to borrow money from Markleeville merchant John Weis to pay for his trip.[31] To keep his men at work during his absence, Chalmers promised them shares of stock in lieu of wages.[32]

He returned to Alpine on August 23 with enough cash in his pockets to begin paying outstanding bills. Far more significant, he'd managed to inspire confidence again among the investors, receiving promises that additional funds would be forthcoming.[33] Even those expected infusions wouldn't be quite enough, but Chalmers was determined to keep trying. This was, he felt, "my last chance."[34]

Things at the mill had not gone well in his absence. Battery samples had been showing $60 to $70 per ton when he left, but "[b]y the time I got back from London our ore had got down to $20 and $28, and our melting operations were damnified by frozen pipes bursting and frozen solutions, pumps bursting, &c &c," he would later recall.[35]

Worse yet, the mine's creditors—aware that the trip had been at least partially successful—hounded him harder than ever upon his return. "[A]s I predicted, everybody therefore expects full and immediate payment," he complained to Syme.[36]

Hopeful that high-level networking might bolster his efforts, Chalmers wrote to Carson City lawyer A. C. Ellis in September 1882, boldly requesting a personal introduction to John Mackay, "the Bonanza King of the Comstock."[37] He pressed the same request on William Lane Booker, the British Consul in San Francisco, reminding Booker that Chalmers had presented letters of introduction from London some fourteen years earlier.[38] There is no indication, however, that an introduction to the Comstock luminary was ever made.

Chalmers did manage to improve his refining technique. By late December 1882, he could proudly report that a grand total of $20,604 had been produced from the Morning Star during the past six months in bullion and cement copper.[39] He was forced to admit, however, that the mine's "rich schute" had been exhausted while he was gone in London. But he assured Syme the ore would again improve if they just kept going deeper. Only 100 feet below the current tunnel, he insisted, was "very rich ore" he had seen with his own eyes eight years earlier. The newly moved hoisting works were finally in place to help raise ore and waste rock to the surface.[40]

"If you expect to see me again in London unsuccessful, you are much mistaken," he informed Syme on Christmas Eve, adding with bravado: "If the mine does give in [fail], I shall not break my heart, for I would rather live on a crust of bread than be worried by such stuff as you write. I trust, however, that the Mine will not fail me."[41]

By early January 1883, the mill's working capacity had been successfully boosted to ten tons per day, and crushing recommenced. The shaft was down to 214 feet, putting them "within 14 feet of a very rich ore body." Once that was reached, Chalmers boldly predicted, "our shipments of bullion alone will average over $400 a day, besides copper."[42]

But by the following month, it was clear that his grand plan to double the mill's capacity was a financial failure:

> My ten tons are not showing a profit, but a loss, because now that our ore carries 11% copper, we find our slime tanks too small to take the charges from the large agitators quick enough; the consequence being that the iron in the precipatory tanks is so covered with slime that it don't act on the salt solution and the solution is pumped back carrying copper, which gives us base bullion. But I shall have this cured next week, when we shall, I hope, be able to show a clear profit.[43]

That was not the only disappointment. Freezing weather had burst pipes in the mill and halted ore hauling.[44] Even with those setbacks, however, Chalmers was still managing to produce copper at the rate of $80 per day.[45]

London had been promising him a remittance of $5,000, but the money had not shown up. "I am utterly impecunious," Chalmers told an attorney representing an unpaid teamster. "[But] in six months I shall be in funds. Press me now, and you will not only ruin me but the Isabelle Company … who have just commenced running their Mill."[46]

In early spring, as his financial condition continued to deteriorate, Chalmers was reduced to trying to borrow funds to get by. He wrote a London friend requesting a $300 loan, "but would like to get $400 if I could," emphasizing that "the Company now owe me $8,000 and over."[47] In late February, he mortgaged his furniture to raise $500 to stave off creditors. "Feeding your miners is getting me into a nice fix," Chalmers groused to Syme. "All my energies are now being devoted to make this thing pay. I expect no aid and will hold on till I burst."[48]

The company was urging Chalmers to again come home, stressing that another in-person visit could raise more money to pay bills and sink shafts.[49] Chalmers appeared to be contemplating it. If nothing else, word of another trip home would soothe irate creditors. Seizing on that for negotiating leverage, Chalmers notified the Vulcan Company that he would do his best to pay their outstanding bills provided they took no steps to file an attachment pending his return from abroad.[50]

When Chalmers talked about mortgaging his furniture, this was probably one of the pieces he had in mind. Today, this velvet-upholstered sofa and matching chairs can be seen at the Alpine County Museum. (*Dustman photo*)

The financial strain was taking its toll. "This debt incubus is killing," Chalmers confided to Syme. "I can neither eat nor sleep, and as sure as you are a living man, we shall be attached and [the work] stopped unless I can stave them off by stating that I am going to England to raise money to pay up."[51]

Bleak Prospects and a Bonanza

Dangling the hope of another trip didn't work. On March 30, 1883, the *Genoa Weekly Courier* carried the humiliating report that "L. Chalmers has had all his property attached."[52] The Vulcan Company, out $8,700 for powder supplied to the tunnel project, had attached the Morning Star's assets. It was, as Chalmers put it, "the fatal catastrophe."[53]

By mid-April 1883, both the mine and the mill were shuttered, with a "keeper" placed in charge. Other creditors, too, had begun filing their own attachments. "[N]o money in the bank, all attached again by [teamster] Mayo and others ... and all black, indeed," Chalmers pronounced wearily.[54]

Antoinette was still not completely well and, "with a helpless baby at her foot," was back to doing the cooking, cleaning and washing. "Have tried to sell my wife's piano and a portion of the furniture to raise a little money for necessaries, but can get no buyer," Chalmers wrote in May, describing his "miserable plight" to Syme. "Sold my rifle to pay my taxes to prevent the sheriff taking my furniture for that purpose."[55]

To friend Wheeler he confided: "Cannot hold out here much longer. No meat in the house for a week, and our staple of potatoes and salt fish about out. I have come through a good many hardships, but have never been in the fix I am in now."[56]

Outright threats were now being made to Chalmers's personal safety, though he tossed them off with his customary aplomb:

> Coming from the Mine on Monday I was met by a miner and informed that if I did not leave Alpine soon, I would be made to do so. When I told him that it would take a good many such men as he to make me go 'till I was ready, he told me to sell my furniture (wife's piano specially) and give him and others money or it would be worse for me. At the mill, my stableman overheard a plot concocted to burn my house after locking me in. (But I was ready for them.)[57]

Hopeful that positive words from the right source could bolster confidence abroad, Chalmers penned a letter to capitalist George Hearst in June, 1883, soliciting Hearst's opinion of the Morning Star Mine. Chalmers's letter hinted at what he hoped the wealthy investor would say: "that you had some years ago examined [it] ... and that you had expressed yourself pleased with same." While he was begging a favor, Chalmers also asked for Hearst's opinion of the Colorado, which Chalmers hoped his friends would purchase, reminding Hearst he had once "examined it for $45,000." While Ashburner and friends had "wet blanketed this district," Chalmers admitted, "I have faith in it yet."[58]

Despite such optimistic declarations, Chalmers's confidence in the Alpine mines was waning. He wrote the owner of the Drum Runner Mine in Montana's Marysville Mining District, asking: "Have you any room for me at a moderate salary? I think I could be useful to you and I am not afraid of work."[59] If he received a response, however, it clearly was not positive; that was the last of his correspondence with the Drum Runner.

Lawyers confirmed to Chalmers that the only way out was insolvency proceedings. Chalmers managed to get work restarted at both the mine and mill by having friend John Weis appointed as a "friendly interim receiver." Weis, in turn, named Chalmers as acting manager. To keep the men at their stations, Chalmers agreed that any proceeds would go first to pay the miners' salaries.[60]

But when $1,200 worth of copper was shipped off in early June, it was promptly attached at the freighter's Nevada ranch, forcing Chalmers to pay off the creditor's claim. To avoid similar surprises going forward, Chalmers concocted an artful dodge. Future copper shipments would be camouflaged as a "purchase" by the freighting company, he advised, complete with fictitious Bill of Sale, "and you will therefore take the shipping receipt in your own name, not in that of the firm."[61]

By July, crews were finding "patches of good ore" in a drift at the bottom of the Morning Star shaft, now 220 feet below ground. Some 70 tons of ore were already at the mill, waiting to be processed. It sounded encouraging. But the "cream of the schute" had already been removed.

To friend Wheeler, Chalmers wrote: "So long as there is a plank in the ship I shall stay by her, but the planks are getting very few."[62] Even more candidly, he advised Syme "the mine is not looking well."

> I am much afraid, friend Syme, it is all up. Of one thing I am certain: I can't possibly stand it much longer. Mrs. Chalmers [is] sick from hard work and getting short of necessary clothing for herself and baby.... Your chances for raising money are evidently very slim, and if I don't run [the mill] soon, the miners will leave and [it will be] another crisis to meet.[63]

Then, in August, a "bonanza" was struck at the 220 level of the Morning Star, due south from the winze. "I am afraid to say too much about my find, [as] I have been so often fooled by the d[amne]d mines, but this time I do think I am cutting into a veritable bonanza," he wrote Syme. "One thing, of which I was sure before [but am] doubly sure now, [is that] your rich ore schute goes down, and that is all I care to know at present."[64]

It was not an immediate success, of course—far from it. In fact, Chalmers was able to sink just 7 feet on this new ledge before "water drove me out."[65] Further work would require not only pumping but a "heavy cash outlay" for timbers, and Chalmers was already having a difficult time finding men to work. Prospective employees were fearful they wouldn't be paid, and another lien sale of the mines was set for mid-September.[66]

Nevertheless, a short run on the ore extracted between August 8 and October 1, 1883, produced encouraging returns: $4,544 in silver; $643 in gold, and more than $4,000 in copper—a total of over $9,200 in much-needed funds.[67]

This happy news prompted London to promise that another £20,000 sterling would be raised to fund continued exploration. The bad news, however: Chalmers could not count on that until November, when the "business season commences" and wealthy investors returned to London from their annual hunts.[68]

The southern end of the Morning Star mine near Monitor, in 2016. First opened in 1863, the Morning Star was originally driven southward from the north side of the hill, eventually coming out here, near Forest Flat. This section of the mine was later known as the Georgianna. (*Photo courtesy of Ed Rogers*)

Faith Falters

Adding to Chalmers's difficulties were troubles with his family. Antoinette was again unwell, "so sick I must send her to San Francisco to be under the doctor for a lengthened period," as Chalmers explained to a mining friend. With only promises from London at the moment, Chalmers buttonholed two local contacts to ask for a loan of $300 or $500.[69]

Son Johnny, having given up his priestly aspirations, arrived at Silver Creek on the last day of September. That meant one more mouth to be fed in the Chalmers household. Now it was up to Chalmers to help find his son a job.[70]

At twenty-three years old, Johnny was "a first-class German scholar, a good bookkeeper, and carrying with him first class testimonials, steady, gentlemanly and willing." His father wrote letters on his behalf, trying to secure Johnny a position with a borax works near Independence, California. But that opening was quickly filled.[71] For the time being, the talented German scholar was put to work helping Antoinette with the cooking, cleaning, and washing.[72]

Meanwhile, back in Eichstadt, daughter Elleny had fallen ill and, in October 1883, passed away.[73] Chalmers urged Jessie, his only remaining daughter, to leave that "death-dealing" climate and "come to your father and second mother." But he held out little genuine hope he would ever see her again. As he wrote sadly to Syme, "Elleny is dead—another victim to Catholicism, and Jessie will soon follow."[74]

Alick had also expressed interest in coming to live with his father. But by now he was married, and his father politely refused. Chalmers's own financial situation was too rocky, and the household already taxed Antoinette's strength and patience. "In these circumstances, much though it pains me, we must deny ourselves the pleasure of having you and your better half here at this time," he wrote tactfully.[75] Instead, Chalmers advised Alick to go to San Francisco to seek work, and sent him a letter of introduction "to the great engineer Von Schmidt." By the look of things, Chalmers confessed wryly, he himself would "soon be on the same hunt for employment."[76]

Adding insult to those personal injuries, Chalmers's groom had "decamped ... in the middle of the night" with his horses, wagon, and harness. Chalmers had given him a bill of sale covering those items as security for unpaid wages. Rather than waiting for his money, however, the groom simply took them and departed, leaving Chalmers without so much as a saddle. "[S]o I have now to walk to the mine and it is deuced hard walking," he wrote to Syme in exasperation.[77] He would soon discover that his "entire stable outfit" had been sold to teamster Dick Gelatt in Genoa, forcing Chalmers to hire back his own horse, saddle and tack for a monthly fee.[78]

A special shareholders' meeting to try to raise the needed funds had been scheduled for 1 November. The plan was to seek permission to issue both new shares and new debentures, which could yield £25,000 for the Morning Star and another £25,000 for the Isabelle Tunnel.[79] In preparation for this crucial meeting, Chalmers persuaded mining expert Charles H. Aaron to inspect both mines—promising him $500 for his report, though it's unclear if Aaron ever got the money.[80]

Aaron was, as Chalmers put it, the "boss metallurgist of San Francisco" and though a "notorious bear" on mining prospects, the report he submitted was largely encouraging.[81] Several shareholders remained unconvinced, reading

Aaron's report as "doubtful, especially as to [the] probable permanence" of the lodes. Nevertheless, the twin funding resolutions passed.[82] That didn't mean money was immediately on the way, however; the company still needed to find investors who would purchase those new shares and debentures.

Two days before Christmas 1883, Chalmers sat down at his desk and penned a gloomy letter to Syme:

> I am much afraid, Henry Syme, it is all up and that our fifteen years of hard work has been time thrown away, leaving us that much older and me, at least, with but a poor chance of anything but a bare existence, if that, and nose at the grindstone for the few years that may remain.[83]

He was nearly right. Chalmers's labors in Alpine were not quite at an end, but that time was growing close.

By mid-January 1884, Chalmers confided to friend Arthur Wheeler, "I am much afraid we shall never start [up] again."[84] Another $4,000 would be required to redeem the mines from the last creditors' sale, and the company's right of redemption would expire in May.[85] More liens were only a matter of time. Chalmers confided to Syme he was "daily expecting an attachment on the house."[86]

Son Johnny had departed for San Francisco "to take his chances." His brother, Alick, was already there, and Chalmers was hopeful both boys would "get something soon" through letters of introduction he had provided them.[87] Antoinette was "killing herself cooking, washing, and scrubbing," and had begun to suffer fainting spells. "She cannot stand it, but does not complain when well," he wrote Syme.[88]

Chalmers began liquidating all the personal assets he could spare. After borrowing a much-needed $100 from merchant John Weis to bail him "out of a scrape," Chalmers bundled up Antoinette's "crockery" set and his own "rifle and glass" and sent them to Weis, asking him to "raffle" them for $50 or $60 "or anything you can get." Chalmers even offered to help sell tickets himself.[89] "I feel degraded by having to live on him [Weis]," Chalmers confessed to Syme, "and will not much longer. I cannot endure it."[90]

Even Syme was not writing as regularly as he had in the past. "Nothing from you since yours of December 16th," Chalmers grumbled on January 26, 1884. "Have you thrown up the sponge?"[91]

Yet despite all his agonizing over money and family matters, Chalmers found time to indulge in his wood-turning hobby. He directed a letter to a lathe manufacturer on January 27, 1884. After complaining about a lathe he had purchased from the company some time ago, he asked "on what terms you would exchange my lathe for one of your most improved foot lathes.... As I am going in for ornamental turning, I would like a Parisian plate of gunmetal on the face of the large pulley and [a] click-spring to catch into the holes."[92]

That complaining letter, as luck would have it, was the final page of Chalmers's bound personal letterbooks—and nearly the last of his letters to survive. With just a few brief exceptions, our glimpse of Chalmers's life through his own words fades away after January 1884. But that is not quite the end of Chalmers's saga—or Antoinette's. Luckily for us, bits and pieces of information from other sources have helped fill in the final chapter of their story.

12

The Final Years and Chalmers's Legacy (1884–1913)

Chalmers dropped a brief note to the Exchequer's directors in early January 1884, reminding them that almost a year had expired since their mines were sold for unpaid taxes, and the redemption window was closing soon. Unless $525 was paid by February 25 to redeem the property from its tax foreclosure, the Exchequer properties would be "gone irretrievably." "Better raise enough to pay all the Taxes," he prodded. "The engine in the Mill alone is worth the money."[1]

While he was at it, Chalmers wove in a polite request for the debentures that had been promised to him in lieu of unpaid salary and expenses. The company's debt to Chalmers now totaled $27,350, with another $755.41 owed him as reimbursement for a personal note he had signed "to save an attachment [by freighter D. W. Earl] of the Isabelle monies in my name."[2]

There is no record that Chalmers received either the cash or debentures he was due. But the directors did heed his request for the considerably smaller amount for back taxes. In May 1884, the Exchequer properties were successfully redeemed from a Mrs. Green, and title officially returned to two of the London directors.[3]

It is there that Chalmers's own voice leaves us. The last entries in his personal letterbook are from late January 1884, and only a few later letters have survived. But from other sources we know that, by hook or by crook, Chalmers managed to keep his Alpine efforts going for another full year and a half.

It was anything but smooth sailing. Even before Mrs. Green was repaid for the 1882–83 taxes, another tax sale took place on 5 February 1884 for Exchequer's delinquent 1883–84 taxes. Once again, Mrs. Green was the successful bidder, paying $360.55.[4] The company still had another year's redemption period within which to buy its property back. But the Exchequer was hanging on by a thread.

As spring gave way to summer, Chalmers received another dose of tragic news: Jessie, his last surviving daughter, died in July 1884 at Eichstadt.[5] Although no cause of death was provided in her death notice, twenty-four-year-old Jessie may well have succumbed to tuberculosis like her sister and brother.

State and local taxes came due again in November 1884, with a fresh tax-default sale looming in February. Chalmers relayed that depressing warning to London but, for once, expressed no particular urgency about it.[6] His bigger concern was

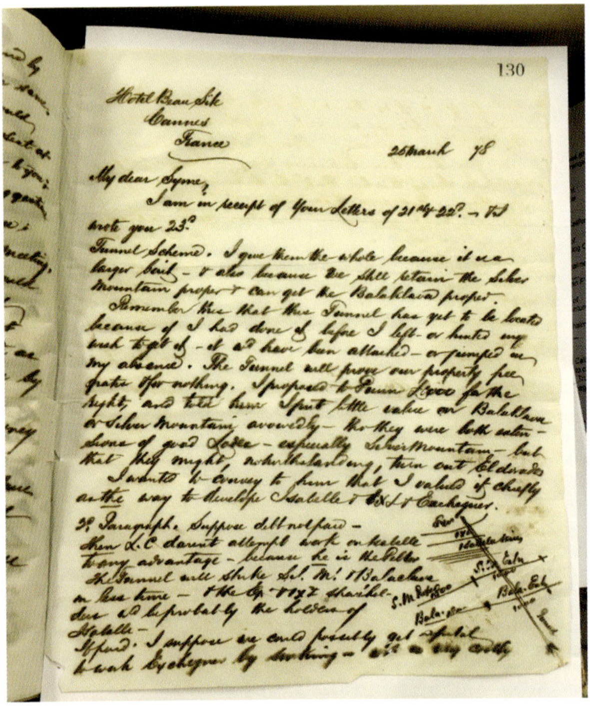
130

Hotel Beau Site
Cannes
France

25 March 78

My dear Syme,

A page from Chalmers's 1878 personal letter book, describing his "Tunnel Scheme" as "the way to develop" the Isabelle, I.X.L., and Exchequer lodes. His hopeful sketch at lower right shows the proposed tunnel striking the Balaclava and Silver Mountain lodes before continuing on to cut the Isabelle, I.X.L. and Exchequer. (*Dustman photo*)

the end of the redemption period from the previous tax sale to Mrs. Green; the company had only until March 1885 to redeem its properties, or lose them forever. As New Year's Day 1885 dawned, yet another worry was added to Chalmers's growing list: Antoinette was expecting again.

That February, Chalmers dutifully compiled all of the Exchequer's accounts from 1877 through 1883 for his London employers—although the mines officially remained in Mrs. Green's hands.[7] With only $360 needed to redeem the property, how could the company not send the funds to reclaim their properties once more? But another last-minute financial bailout was not in the cards. On March 7, 1885, the redemption clock having ticked through its last day, Mrs. Hannah Green became the official owner of the Exchequer mines, including the Stella/Morning Star.[8]

If there was one bit of comfort for Chalmers amid all this gloom, it was that his son, Alick, was living nearby. A gifted oil painter in addition to his talents as a surveyor/draftsman, Alick had been commissioned by Judge Arnot to paint several local scenes. Among the works Alick produced in rapid succession were a painting of Silver Mountain City, finished in May 1885; a street-scape of Markleeville, featuring both the wooden courthouse and Judge Arnot's home, completed in June; and a view of the I.X.L. Mill, said to be the "finest work of the three," finished in July.[9, 10, 11] Alick's work was well-received; by September he would be engaged in painting the nearby town of Woodfords for its prominent storekeeper, W. P. Merrill.[12]

Alick's wife, the former Annie Banks, arrived in Markleeville that April for a visit, then ventured on to Silver Creek to see her in-laws, including the very-

Peter Curtz, a fellow miner, was a good friend to Chalmers, possibly even saving him from self-harm at one point, and Curtz's family would later warmly embrace Chalmers's daughter, Laura. Curtz is shown (second from right), with his wife and family inside the Curtz Mill near Bulliona, California, *circa* 1890. (*Alpine County Historical Society*)

pregnant Antoinette.[13] Just six weeks after Annie's visit, Antoinette gave birth to her third child: a daughter, born at home on May 30 1885.[14] The new baby was named Laura, likely in remembrance of Chalmers's older daughter who had died three years before.

On June 29, 1885, Chalmers sat down to compose his last official letter for the Exchequer Company, a mundane bit of correspondence advising the company's lawyer of the fees for recording deeds in Alpine County: twenty cents per page for the document itself, plus another twenty-five cents for indexing the document.[15] Now with a wife and three children to support and no reliable source of income, Chalmers must have felt desperate indeed. He renewed his effort to sell Antoinette's piano, a Grand Square Antisell in "good as new condition," advertising it for half its original $500 purchase price.[16]

Locals would later say that Chalmers's friend, Monitor mining entrepreneur Peter Curtz, paid a fortuitous visit to Silver Creek about this time, arriving just in time to keep the "despondent" Chalmers from ending his troubles with a pistol.[17]

Bidding Silver Creek Adieu

It is difficult to know what prompted Chalmers to undertake one more trip to London. Perhaps he hoped that his personal charm, applied liberally and in person, would once again start a fresh flow of funds. Perhaps he had simply had his fill of mining. Or perhaps he had had enough of the high-strung Antoinette.

Whatever his reasoning, Chalmers made up his mind to leave. On August 31, 1885, three days before his departure, Chalmers sold 160 acres of timber land across the road from his house to employee Dominick Bari, for $200—likely to raise cash for his trip.[18]

On September 3, 1885, Chalmers was seen passing through Genoa on his way back to London. The *Genoa Weekly Courier* put the best possible spin on his plans, reporting that he was returning "in search of more capital with which to push work in the mines of Alpine County." A news note the following week predicted "prosperous times" if Chalmers's efforts were successful, adding optimistically that he expected to be gone only "about three months."[19]

Left behind in the house at Silver Creek were Antoinette and her three children—including three-month-old baby, Laura. But Chalmers had taken steps to ensure Antoinette was not entirely alone. He left her under the care of the same trusted employee to whom he had sold the land—Dominick Bari.

A thirty-eight-year-old Italian miner who had worked for Chalmers for several years, Bari had been living at the nearby boarding house. But he soon moved in with Antoinette as a boarder, putting him closer for help with household chores. He would continue to correspond with Chalmers on business matters for several years. Antoinette likely welcomed both the assistance and the company. She was now thirty-two years old—far closer to Dominick in age than to her sixty-year-old husband.

Chalmers's reception when he returned to London is unclear, but with his dismal track record, he probably was not welcomed nearly as warmly by his friends and investors as he had been in the past. One small clue to his efforts appeared in the *Aberdeen Journal* roughly nine months after his return: an abridged prospectus

The only known photograph of Dominick Bari, Antoinette Chalmers's long-time companion. (*Courtesy of the Thornburg family*)

for the Colorado No. 2, touting Chalmers's glowing report on the mine.[20] True to form, it seems he was attempting to float yet another new mining company in hopes it would resurrect his battered fortunes.

It is hardly surprising that Chalmers would turn his attention to the Colorado No. 2. He had had his eye on the mine for quite some time, and its Tarshish lode had produced significant quantities of silver in the past, making it a better bet than the numerous untested claims Chalmers had pursued in the past.[21] In what likely was a related development, son Alick was dispatched to Bulliona in September 1886 to survey several mining claims.[22]

Nothing came of that push, at least immediately. But there were hints for a while that the gamble might actually pay off. A full year later, rumors were circulating in mining circles that Chalmers "is expected to arrive from England, with capital to resume operations on the Colorado No. 2 mine."[23]

It was apparently just a rumor, however. The Colorado No. 2 would change ownership several times in the coming decade, eventually winding up as part of Peter Curtz's Monitor Mining & Milling Co.[24]

Chalmers continued attempting to gin up interest in other Alpine mining adventures where he could. In late 1888, a flurry of British attention was directed toward the old Morning Star mine. Londoner William Gunn purchased the Morning Star from Hannah Green for a hefty $2,000—a price reflecting serious interest. Green had paid just $360 for it four years earlier.[25]

The sale of the Morning Star helped fuel another spate of rumors about Chalmers's imminent return in mid-1889. "It is probable work will be resumed on the mines of Alpine County," enthused Carson City's *Daily Appeal* that July, "as it is reported that Lewis Chalmers is being backed by English capital for another try in that mountainous region."[26] A similar report appeared in the *Bridgeport Chronicle* that same month, announcing Chalmers "has made another raise, and is backed by a company that will commence active operations again in the county as soon as Mr. Chalmers returns."[27] Three months later, the *Genoa Weekly Courier* echoed the sentiment: "The indefatigable Chalmers is said to be at it again with a London sack," it claimed, and "an effort is being made to revive mining business in Alpine."[28]

The ever-sanguine Chalmers himself was almost certainly the source of all those rumors. But his dismal track record seems to have finally caught up with him. Though there may have been a few sparks of interest, none of them ever took fire and Chalmers himself would never set foot in Alpine again.

Time Rolls On

Meanwhile, time and the elements were taking their toll on the products of Chalmers's labors. His two-story house at Bulliona, built during his earliest days at the Imperial Tunnel, was sold off as lumber in 1889 by Judge Arnot, taking one "muley cow" in exchange.[29] The expensive, much-renovated Exchequer Mill at Silver Creek was crushed in by heavy snow in January 1890.[30]

Antoinette was accustomed to harsh winters at Silver Creek, with 3–5 feet of snow. But the winter of 1889–1890 proved truly extraordinary, burying her house beneath 16 feet of the white stuff. Antoinette and her family "were compelled to go out through the second story windows," the *Genoa Weekly Courier* reported. Lights inside the home had to be kept burning all day long, "as the snow is piled up so high that it entirely shuts out the sunlight."[31] Silver Mountain City by now was virtually a ghost town, and Antoinette became even more isolated in November 1890, when mail service to Silver Creek was discontinued. Thereafter, she was forced to pick up her mail in Markleeville, 8 miles away.[32]

Chalmers continued to send occasional letters, but remained in London. The 1891 census found him a boarder in the home of widow Constance de Ridder, at 77 Navarino Road, Dalston, Hackney, sharing a household with four of Ridder's children and her son-in-law. It was a respectable though not wealthy neighborhood. In a bit of wishful thinking, Chalmers listed his occupation as "mines manager."[33] In reality, however, the highly-educated Chalmers may have been working as a humble bookkeeper or company clerk to support himself.[34]

Chalmers hadn't given up entirely on the Alpine mines' potential. Fresh local rumors began circulating in March 1891, hinting that an "English company" was prepared to spend $10,000 to reopen an old tunnel, with another $240,000 to arrive by summer. Having been stung on such glowing predictions before, however, the *Genoa Courier* cautiously hedged its reporting: "Just how much importance is to be attached to this statement we are unable to say, but some of the residents of Alpine confidently look for a revival of the mining interests."[35]

While the mine in question was not identified, this may have been an attempt to resurrect the Morning Star. William Gunn's deed (signed three years earlier) was recorded in April 1891, just a month after the *Courier* ran its story.[36] There may also have been renewed interest in the Scandinavian Canyon mines. That same summer, mining patents were recorded for a cluster of claims in Chalmers's name.[37]

Silver Mountain City was practically a ghost town by 1890, when mail service was discontinued. (*Photo circa 1885, Alpine County Historical Society*)

Fresh tragedy struck in May 1891, while Chalmers was still in London. Little Lewis, his nine-year-old son with Antoinette, drowned in the fast-moving waters of Silver Creek behind their house. The child was laid to rest in the same small family plot that held the remains of his older half-brother Lewie, who had died of consumption.[38]

Antoinette had no husband at home with whom to share her grief, but she did have Dominick's shoulder to lean on, and the tragedy may have brought them closer. The community also rallied to Antoinette's support. Local ranchers traveled as much as 20 miles to attend the young boy's funeral. One warm-hearted ranch wife even stayed with Antoinette awhile to comfort her.[39] Antoinette placed a "card" in the newspaper, thanking those who had helped search the river for her son's body and "in other ways helped to lighten my heavy burden."[40]

Chalmers likely received this tragic news by telegram; he touched on the tragedy briefly in a business letter to Bari on August 28. Hard up as always for money, Chalmers was trying to arrange a visit to the mines by potential investors, still "hoping to put things straight." Chalmers's grief is evident. But as the letter makes clear, it was Dominick Bari who was now shouldering the burden of looking after Antoinette and her two remaining children:

> I received your letter of July 20th, and write now, under cover to my wife, to thank you for all the kindness and good offices you have shown my family during my enforced absence from them, and particularly during the late severe affliction which has overtaken us [the drowning]....
>
> I am, I need scarcely say, much distressed that I am, as yet, unable to send you the money you so generously advanced to my poor wife. For many years, ever since I arrived here, London has been financially dead, so far as mines are concerned; all my old rich personal friends [now] dead or gone away, and to make a new Connection in the face of all this, and our many failures in Alpine, is not accomplished easily.
>
> But I have not altogether failed in doing this; and in a fortnight or so, a Mr. Dunstan, who has undertaken to form a new Company to take over all these mines from the old Companies (prospectus and reports printed) says he and a Mr. Condrel will leave here to go to the mines via the Big Trees, and I have therefor given them an order to Jack [Dickey] to show them the Mills and Tunnel, telling Jack on the order to ask you to show them the I.X.L., Isabelle, Exchequer and Stella Mines, as you know them (the mines), and Jack doesn't. You need not mention this to Jack until he asks you to go with these gentlemen when they arrive.
>
> In connection with this matter, should the properties be offered for sale at such a small figure as you could afford the money to buy, by all means buy them in, in your own name, but, on no consideration must you allow these gentlemen to know that such a thing is possible, because they might do it themselves, and leave us out in the cold.
>
> Again thanking you and hoping soon to be at the Mill to put matters straight once more, I remain, yours very truly,
>
> Lewis Chalmers[41]

Chalmers threw in a quick, hopeful postscript: "I will write Mrs. Chalmers when those gentlemen are likely to arrive." It was yet another effort, but nothing would come of it.

Brief mentions of Antoinette and her daughter appeared in the newspaper as the years rolled by. In December 1891, six months after her son's drowning, Antoinette brought Laura to Carson City to have her photograph taken, likely to send to Lewis in London. It was the "first time Mrs. Chalmers has been out of the mountains of Alpine for seven years," the paper reported.[42] But that trip seems to have broken her self-imposed isolation. Antoinette and her daughter visited Markleeville a few months later, with Antoinette continuing on to Carson.[43]

Meanwhile, back in London, Chalmers was still dabbling in mining investments. His long-time friend Peter Curtz deeded the North Star Mine (one of the Morning Star claims) to Chalmers for one dollar in May 1891, and Chalmers promptly transferred it (again, for just one dollar) to William Gunn.[44] Chalmers also likely had a hand in an 1892 sale of the Morning Star to Londoner H. H. Seymour, though the deed was not recorded until nearly three years later.[45]

Chalmers was also doing a tiny bit of investing of his own. In 1893, he acquired a small interest in the newly-formed Bordeaux Wharves, Maritime Warehouses and Lands Company, Limited. This company had been formed, it grandly proclaimed, to "acquire land, buildings and hereditaments of any tenure or description in the port of Bordeaux and its neighborhood" and to develop accessory "factories, warehouses, wharves, jetties, piers, docks, quays (etc.)." No small part of its appeal for Chalmers was probably that shares cost only one pound apiece. Chalmers's name appeared in the newspaper among the company's investors, once again identifying his occupation as "mining engineer," with a business address at 72 Bishops Gate Street.[46]

More than a decade after Chalmers shook the dust of Alpine County from his shoes, fresh hints surfaced that he was poised to make a return. In November 1898, the *Genoa Courier* advised its readers that Chalmers had "again secured the aid of English capital to work the mines of Silver Mountain," and "Mr. Chalmers himself is now on his way here from London."[47] A similar tale cropped up six months later, assuring readers that the "English company formerly represented by Mr. Chalmers" was ready to hire, and predicting "work will be resumed soon on the Stella mine."[48]

The Dream Unravels

Despite such hopeful talk, the only change on the ground was a slow dismantling of Chalmers's empire as all the fruits of his labor were scattered. The machinery at the nearly-new I.X.L. Mill was removed about 1896 and carted off to Mono County to work ore at the Dunderberg Mine.[49] Sadder yet, the improvements at the Exchequer Mill compound—including Chalmers's "mansion"—were sold at public auction in February 1899 to satisfy a $1,200 lien by Judge Arnot. Once again, a year-long redemption period applied. But when that expired the following year, Arnot received a sheriff's deed to it all: the moldering eighteen-stamp quartz mill, long-vacant boarding house, Chalmers's assay office, and his house.[50]

Antoinette, now forty-seven, was still living in that house, with twenty-five-year-old son Henry and fifteen-year-old daughter Laura. Dominick Bari still lived with them, and Antoinette had taken in another boarder to help meet expenses.[51]

That presented a thorny problem for Judge Arnot. The property was now his, but he couldn't in good conscience evict Antoinette and her family. In the end, Arnot carved out a five-acre parcel of land for the house and its two fenced gardens, and quitclaimed it to Antoinette in July. He carefully wrote in an exception: the expensive Exchequer Mill machinery was not part of the deal. Arnot also did not give Antoinette the house for free. She paid $105 for it.

It was Bari who submitted the quitclaim from Arnot for recording—a minor courtesy that suggests Bari might have helped negotiate this solution. He had apparently been assisting Antoinette with business matters since Chalmers's departure fifteen years earlier.[52] Bari also became something of a father-figure to Laura as the years went by. According to apocryphal stories, he gave her a flock of lambs to raise since she had no playmates nearby.

Antoinette eventually sent Laura to live with the Ellis family in Diamond Valley, so she could attend school. [53] But Laura's time with the Ellis family proved an unhappy arrangement.[54] She tried to run away, giving the Ellises quite a scare, and once scrawled a cryptic note threatening to kill herself. It was clearly a cry for help, and Laura was ultimately sent to live with the Curtz family in Oakland, California.[55]

Lewis Chalmers may indeed have had aspirations of returning to resurrect his mines someday. But he never did see his wife or daughter again. For those of us who sense a kindly, gentle soul behind his high-flying dreams, it is hard to reconcile this seeming lack of concern. Why would he not have sought another job in the States where his family could be with him, or arrange for them to join him London? Was he truly that cold-hearted or were other factors at play?

Perhaps Chalmers was not eager to face his own bitter disappointments. In the anonymity of London, he could pretend (to himself, at least) that he was still a gentleman, and his fortunes would eventually turn around. It was also far easier to eke out a meager living in London than in rural Alpine.

It is also possible that he had had his fill of living with Antoinette. She could certainly be needy and difficult. Contemporaries who knew her described Antoinette as "odd," hinting that she suffered from a mental disorder. To someone like Chalmers who fancied himself an upright and honorable man, long-term separation may have seemed preferable to an actual divorce.

If Chalmers had tired of Antoinette, however, the feeling was not mutual. Antoinette missed him terribly and remained convinced that Lewis would return to her someday. She kept his office exactly as he had left it. "[O]n his desk reposed his old Meerschaum pipe, where he had laid it after enjoying his last smoke there, with the ashes half spilled out on the booting pad," one Alpine native would later recall.[56] For years, Antoinette would walk up to the stage stop on the road near her house whenever the stage was due to pass by, hoping Chalmers would be on it. But after years of disappointment, she eventually lost hope and "was seldom heard to mention his name."[57]

London's 1901 Census shows Chalmers still boarding with the family of Constance Ridder in Dalston, Hackney, now at 250 Richmond Road.[58] He would mourn the death of yet another son that year: Alex Chalmers died of sunstroke in June 1901 at the age of forty-seven. He had remained in Nevada after his father's departure, becoming a "well known civil engineer." On the day of his death, Alex had been surveying in Spanish Springs Valley, Nevada. He was buried at Hillside Cemetery in Reno.[59] His survey work had not been limited to the Reno area; he had surveyed "nearly all the ditches in [Carson] valley," spent time in Churchill County, and mapped waterways near Genoa.[60,61]

In mid-1903, another push was made to resurrect the Alpine mines, though Chalmers does not seem to have been behind it this time. Instead, this fresh burst of enthusiasm was sparked by news that Peter Curtz planned to reopen both the Colorado No. 2 and Morning Star mines. Though nothing much seems to have come of it, optimism ran high for a short time, with the *Gardnerville Record* gushing that "the old camp of Silver Mountain will rise Phoenix-like from the ashes of the ruins that now abound." The "old I.X.L. and Exchequer Mining ground" were still in Judge Arnot's hands, the reporter observed, as if that alone was a predictor of success. And, at least in the reporter's estimation, the 4,000-foot Isabelle Tunnel remained "one of the greatest mining tunnels ... that has ever been worked out in this or any other section." As for the mines' previous failures, well, those had stemmed simply from "mismanaged funds and lack of good reason."[62]

A similar column the following week lauded the I.X.L. and Exchequer mines, noting that "considerable bullion" had been extracted until water in the vertical shaft forced Chalmers to switch course and pursue his horizontal tunnel instead. But the piece bore the title "Silent Monuments of British Folly," as if the battle were already lost, and the writer spared no words of sympathy for Chalmers:

> [He] deserted his wife and babies, and left them to struggle with the fast-decaying camp. He never returned and today his wife remains in one of the houses that marked his extravagance in the early days.[63]

Eighteen years had now passed since the last workman threw down his tools and departed for greener pastures. The mines and their structures were decaying, sad remnants of once-lofty dreams. Old ore cars, rusting Burleigh drills, abandoned blowers and other equipment lay scattered at the mines.[64] The Isabelle Tunnel's stout, brick-lined powder magazine remained intact, though its outer door had been rudely ripped away. Some 15 tons of Giant Powder still sat inside, with glycerine now seeping out of the powder.

> It is said that Chalmers spent over a million and a half dollars in the prosecution of the work, and only left a monument to his folly.... [T]housands of dollars worth of tools were left in the tunnel or cast aside when the final breaking up came ... and the camp in a few months was as deserted as it is today.... There was enough English money used in this section to carry on a war for several months and the worst of it all, the Englishmen received no bullion values in return.... all [now] monuments to a folly of men, who mined without knowledge of what they were after, and monuments to the wealth of British money advances.[65]

On February 21, 1904, the ever-hopeful Lewis Chalmers finally reached the end of his long struggle. He passed away in London's Hackney district of "heart complaint" following a short illness. Had he lived another sixteen days, he would have celebrated his seventy-ninth birthday.[66] His obituary in the *Reno Gazette* remembered Chalmers warmly as a "generous, high-minded English gentleman," reporting he had sunk an astonishing $1.5 million in British funds into his mills, furnaces and tunnels. Unfortunately, the writer concluded, he "knew nothing of mining."[67]

The *Record-Courier* was considerably less charitable, describing Chalmers's legacy as "one of the most expensive tunnels ever attempted in the west":

> Handsome homes were built, a foundry constructed, saw mills were put to work and a city started. During the time that the tunnel was being driven he had two mills erected to reduce the ore when they came upon it.... The entire plant of Chalmers was one of excessive extravagance. Had he confined himself to driving an ordinary tunnel he would have landed at the objective point and the existence of the ledge would have been told.[68]

Saddest of all, in the eyes of sympathetic locals, was the result for Antoinette:

> He left his wife in the deserted village. She alone had to face the disappointed miners and people. She has struggled in poverty ever since he left, and year after year she has eked out an existence, to the surprise of the people of Alpine county. The disappointments that were showered upon her and the poverty have partially unsettled her mind.... [A]s long as the camp retains its name and as long as the brick foundations remain there will be a memory of the man who squandered money like water, and forgot the woman whom he called his wife.[69]

Antoinette continued to share the old house at Silver Creek with Dominick Bari, but it was an isolated existence. Always somewhat odd, Antoinette grew odder yet following her husband's death. Locals would later claim she answered the door with a vial of acid in one hand and a pistol tucked in her apron: the acid to ward off ghosts, and the pistol to deter human intruders.[70]

She still had her garden, and Dominick would have kept the house supplied with firewood. The community sympathized with the widow's plight, and she may have received occasional assistance through neighbors' charity. But it was undoubtedly a tough existence. Daughter Laura had reportedly married an Oakland bookseller named Carsten, but did not keep in touch.[71] Son Henry Laughton, now grown, was living in Markleeville.[72]

By 1913, Antoinette's situation had become even more precarious. She was now sixty years old and increasingly eccentric. Local law enforcement insisted it was too dangerous for her to spend another winter in the isolated house, pressuring her to leave and "go into a rest home." For a short while, Antoinette moved to the Alpine Hotel in Markleeville, living with the Ellis family.[73] But that only made her more restless and unhappy. Unable to reach her daughter, Antoinette decided to sell the old house at Silver Creek and go to Oakland to search for her.

Antoinette had become fond of young Marjorie Whitmore, whose family camped near Silver Creek every summer, and determined to sell her the house. But Marjorie was still just a junior in high school.[74] So on November 8, 1913, Antoinette deeded the property to Marjorie's mother, Mrs. Robert K. Whitmore of Ceres, California. The Whitmores paid Antoinette $10, with $400 more due in one year.[75] Antoinette wrote out a handwritten will the same day.

Antoinette packed a trunk with her most precious possessions—including silver dishes and her treasured wedding gown—and departed for Oakland, only to have her trunk stolen along the way.[76] Luckily, she still had a few possessions with her: some jewelry, a mink coat, and a bit of cash.

Reaching Oakland, Antoinette opened a bank account for $370 and rented a room on Piedmont Avenue. But the weeks rolled by with no sign of her daughter Laura. Antoinette spent Christmas alone in a strange city.

Distraught, friendless, and quickly running out of money, Antoinette sought the only peace she could imagine. Three days after Christmas, she turned on the gas jets in her rented room, stuffed paper into the crack under the door, lay down on the bed, and welcomed death.[77] Son Henry arranged for Antoinette's body to be cremated, and her ashes were returned to the family plot near Silver Creek.[78]

Antoinette's holographic will left the bulk of her estate to Henry, her son by her first marriage, with only a token $1 to daughter Laura.[79]

Daughter Laura's disappearance from her mother's life appears to have been deliberate; when the will was probated, even half-brother Henry had no idea where Laura was living.. But Laura did keep in touch with several friends from Alpine County through the years. According to them, she would marry twice more, becoming "Laura Schumacher" and then "Laura Ansty," but eventually spending her final years "as a welfare case in a remodeled garage" in Los Angeles.[80]

Henry Laughton lived the rest of his life in Markleeville, working as a custodian at the Alpine County courthouse. Locals recalled him as something of a recluse who "made the best bread you ever tasted." But he was "very crazy," an acquaintance noted, adding: "he got that from his mom."[81]

Henry fell ill in February 1947 and, after three weeks in the Reno hospital, expired on March 11, 1947, at the age of seventy-two.[82] He was laid to rest near his mother, half-brother, and step-brother in the tiny family plot by Silver Creek.[83]

As for Chalmers's offspring, only two of the remaining seven children from his first wives survived him. Unlike Alex and Johnnie, son Willie remained in England. He married in August 1880 and made his home in Middlesex, England, passing away there at age sixty-seven on November 23, 1923.[84] After giving up the priesthood, younger brother Johnnie married the former Minnie Moore. They lived for many years in Los Angeles, California, and Johnnie died there in July 1940 at almost exactly the same age as his father, one day before his seventy-ninth birthday.[85]

Chalmers's first daughter, Lewisa, who had been born out of wedlock in 1850, also survived him. Lewisa had been brought up in Scotland by her grandparents, going by 'Louisa' after her christening. She married a traveling sales representative named James Blackhall in Aberdeen in 1871, and they emigrated to Toronto, Canada in 1875, raising a total of eight children. Lewisa/Louisa passed away on March 18, 1935, at the age of eighty-four.[86]

Dominick Bari, Antoinette's boarder and companion for more than twenty years, moved to Markleeville, taking a room briefly at the Alpine Hotel. But his health quickly began to worsen. William A. Thornburg and his family took Bari in and cared for him until his passing. Dominick died on August 9, 1914, eight months after Antoinette's suicide, leaving behind a surprisingly large estate that included a $2,730 bank account, $210 in gold coin, and several mining interests. His executor ensured Bari was given a fine send-off; he was buried at the Markleeville Cemetery in a "French Burle casket, fully trimmed" that cost $85, with his resting place marked by a black granite headstone costing another $325.[87]

Above left: The Chalmers family burial plot, not far from their home at Silver Creek. Today's Highway 4 (added later) skirts the edge of the small cemetery. Resting here are the remains of Chalmers's son Lewie from his first marriage, who died of tuberculosis; Lewis, his son with Antoinette, who drowned in the creek; Antoinette Laughton Chalmers; and her son, Henry Laughton. (*Dustman photo*)

Above right: Henry Laughton, Antoinette's son by her first marriage, *circa* 1935. According to local lore, Henry had a wooden leg, the result of a childhood accident, but he supported himself working as a custodian at the Alpine County Courthouse. (*Alpine County Historical Society*)

The Chalmers Legacy Lives On

Witch Hill House in Fraserburgh, Scotland, one of the earliest of Lewis Chalmers's undertakings, was leveled by a tragic fire on February 11, 2012. That was the lavish country house whose construction threw Chalmers into bankruptcy and prompted his decision to seek a career in mining.[88] In some ways, that would be a fittingly catastrophic end to the long, sad tale of Lewis Chalmers. But it is not quite the end of his legacy.

Not all of the mines he pursued in Alpine were utterly worthless, though far more was spent on them than was ever taken out. As much as $70,000 in ore may have been successfully extracted from the Exchequer alone.[89] Chalmers also repeatedly swore that the I.X.L. had produced $50,000 in 1868 alone, an incredible fortune at the time.[90] Chalmers certainly was not alone in failing to find a way to extract Alpine's mineral wealth economically; that goal would elude other later mining capitalists as well.[91]

Even today, tokens remain of Chalmers's time in Alpine County—breadcrumbs, as it were, from his happier days. The old house by Silver Creek is decaying but still standing, and still known as "Chalmers Mansion" by locals. Pieces of Chalmers's expensive birds-eye maple furniture are on display at the Alpine County Museum, and old photos capture how the interior of his house once looked.[92]

Also preserved in the Alpine County Museum are several of Alex Chalmers's original oil paintings from his 1885 visit. Executed with a draftsman's precision, these include an image of the recently-finished I.X.L. Mill, and the only known view of Markleeville's Main Street as it looked before the Great Fire of August 1885.

But the most priceless relic that Chalmers left behind was his letterbooks. Antoinette kept his collection of bound books safe for nearly thirty years after her husband's departure, and they remained in the house at Silver Creek when the Whitmore family bought it. Luckily for later historians, the Whitmores recognized the treasure these documents represented. Robert Whitmore made sure that the volumes came home again, donating them to the Alpine County Museum in 2005.

Part of the fascination of Chalmers's life story is the convoluted path it took. His eventual course was so very different than his upbringing suggested. Yet he managed to survive, and even thrive for a time, at an unfamiliar endeavor in a strange and alien land.

Thank heavens he was so organized, so meticulous, and such a persistent wordsmith. Through Chalmers's detailed correspondence, we get to learn about far more than the eventual failure of his mines. We are treated to a ring-side seat for the dramatic ups and downs of his adventurous life. We feel his passion, hear his voice, and enjoy his "gift of gab." Through his eyes, we see first-hand glimpses of luminaries like Sutro, Hearst, and Von Schmidt.

Through this incredible collection of letters, Chalmers's life, his loves, and his legacy live on.

Above left: Locals still call it the "Chalmers Mansion." The tall stone-and-brick chimney from the long-gone Exchequer Mill is in the foreground. (*Photo courtesy of Jay Topping*)

Above right: Remains of ore rails at one of Chalmers's mines in I.X.L. Canyon. The mine was worked, off and on, as late as the 1960s. (*Dustman photo*)

Endnotes

Chapter 1

1 A "baron bailie" was a Scottish magistrate who acted as the deputy of a baron in civil or criminal matters.
2 *Aberdeen Journal*, October 28, 1840.
3 Henderson, J. A., *History of the Society of Advocates in Aberdeen* (1912 Aberdeen University), pp. 118–119.
4 "Testimonials in Favour of Lewis Chalmers, M.A., Advocate and Notary Public, Late Baron Bailie of Fraserburgh an Justice of the Peace of the County of Aberdeen" (Private printing, 1865, *copy at* Sutro Library, San Francisco, #CT828, C5, T4).
5 Chalmers's illegitimate daughter, Lewisa Chalmers, was born April 30, 1850, to mother Elizabeth Moir of Banff, Aberdeenshire. Her birth registration confirms that the "Kirk Session" of the Banff Kirk had investigated the circumstances of her birth, and was satisfied Lewis Chalmers was the father. Lewisa was brought up by her maternal grandparents, and went by "Louisa" after her christening. She eventually married a 'commercial traveler' named James Blackhall on July 7, 1871, in Aberdeen, and they emigrated in 1875 to Toronto, Canada. The Blackhalls had a total of eight children; one son (James Blackhall, born January 24, 1874) became a Forest Supervisor in Wyoming and helped establish Hayden National Forest. Lewisa/Louisa Chalmers Blackhall died March 18, 1935, at the age of eighty-four. (Genealogical research compiled by family researcher Gordon Harland and by Lewisa's great-nephew, James F. Miles, in possession of the author).
6 Henderson, *op. cit.*, p. 119.
7 Chalmers would later confirm in his letters that first wife Eliza's death was "caused by inflammation (puerperal) after her confinement." Chalmers Letters, *1868–1869 Personals*, p. 13 (March 25, 1868).
8 Information about the estate distribution for Lewis Chalmers, Sr. from family genealogy researcher Gordon Harland (correspondence with the author, July 16, 2010).
9 "Testimonials," *op. cit.*, p. 10.
10 Obituary: "Death of Bailie Chalmers, Late of Fraserburgh," *Fraserburgh Herald*, March 8, 1904; *see also Aberdeen Journal*, January 25, 1860 regarding appointment of Chalmers as Captain of the Fraserbergh Volunteer Artillery Company.
11 Obituary, *Fraserburgh Herald*, *op. cit.*
12 *Ibid.*
13 "Testimonials," *op. cit.*, p. 12.
14 *Ibid.*
15 "Testimonials," *op. cit.*, p. 5. It is unclear whether this may have had anything to do with extended litigation by Harvey Thomsen, husband of Chalmers's sister, involving a marital trust and settlement of the elder Lewis Chalmers's estate. *Aberdeen Journal*, January 18, 1865, and December 1, 1869. Chalmers himself would later allege that "Sir A. Anderson … made off with £4,000" from Chalmers's bankruptcy estate, and that the refusal of Lord S. (likely Lord Saltoun) to pay Chalmers for "Rothen and Witch Hill" had reduced his assets considerably. This suggests that Chalmers was held personally responsible for building costs

he had expected Lord Saltoun to pay. (Chalmers Letters, *1878 Personals*, p. 236 (April 25, 1878)).

16 Letter from George A. Day to Remi Nadeau, November 2, 1995 (in author's possession), recounting a conversation with Dr. David Morrison of Fraserburgh about Chalmers's dismissal by Lord Saltoun. *See also* Lady Saltoun's website, fraserchief.co.uk/places.html for the history of Cairnbulg Castle. The Saltouns did have a history of traveling for extended periods; *see Aberdeen Journal*, October 9, 1861, for example, mentioning that they had left Brighton after a stay of seven weeks for their town residence at Hereford Street.

17 A brief reference in Chalmers's letters alludes to his financial "fiasco," without shedding much light on exactly what that entailed. Friend Henry Syme cautioned Chalmers not to "throw away substance for shadow," warning he was "about to repeat the Fraserburgh fiasco." Chalmers Letters, *1868–1869 Personals*, pp. 96 (July 6, 1868).

18 Henderson, *op. cit.*, p. 119.

19 *Aberdeen Journal*, June 8, 1864 ("Notice to Debtors and Creditors").

20 *Aberdeen Journal*, June 1 and 15, 1864.

21 *Aberdeen Journal*, July 26, 1865.

22 *Aberdeen Journal*, July 27, 1864, and July 19, 1865.

23 Obituary, *Fraserburgh Herald*, *op. cit.*

24 "Testimonials, *op. cit.*, p. 2.

25 *Aberdeen Journal*, April 26, 1865, reporting the arrival of first train.

26 *Aberdeen Journal*, June 21, 1865. *See also supra n. 3* for copy of the "Testimonials" themselves.

27 "Testimonials, *op. cit.*, pp. 8-9.

28 *Aberdeen Journal*, August 31, 1864.

29 "Kelman-Busby Family Tree" (typescript genealogical research compiled by family researcher Gordon Harland, in possession of the author) puts Chalmers in London in November, 1865. This is consistent with the representation in the Testimonials by Johnson, Matthey & Co. dated July 1866, certifying that Chalmers had been studying assaying with them and was now "fully qualified" to assay gold and silver. ("Testimonials," *op. cit.*, p. 12.)

Chalmers's financial difficulties would continue for several years after his move to London, however. A meeting of his creditors was still being noticed in May 1867, for example. (*Aberdeen Journal*, May 15, 1867). Chalmers would also write in frustration to his Trustee in April, 1868, asking about the prospect of an "immediate discharge." (1868–9 Personals, pp. 40, 43, and 45 (April 13, 1868)). A dodgy loan was apparently part of his winding-up strategy; Chalmers would later write that he "had to borrow money to pay my indebtedness before I left London," for which his sister-in-law Mrs. Chalmers served as the "apparent borrower," with himself and brother John signing as "sureties." (Chalmers Letters, *1871–73 Personals*, p. 87 (May 21, 1871)).

30 Prospectus of the Imperial Silver Quarries Company (1867) (Bancroft Library manuscript collection, Item #xF 862.3 I4), pp. 26–27. The prospectus was widely promoted; an advertisement appeared as far away as the *Waterford (Ireland) News*, November 15, 1867.

31 Capt. James Jones claimed to be the "pioneer of the Monitor Mining district," stating that he and his brother Frank were the first to locate claims there; gave the district its name; framed its mining laws; and selected and located its best mining locations. (Michigan University Archives, Jones Collection, "Prospectus of the Active Company" (*circa* September 1864)). Brother Frank seems to have been the true pioneer of the pair, however. In an 1863 letter he describes his living accommodations at Bulliona: "I make my home on the River, [and] have to cook out of doors." (Letter from F.A.S. Jones to his brother James Jones, August 17, 1863, Michigan University Archives, Jones Collection, Folder #1).

32 *Ibid.*, p. 3 (listing the Imperial's directors). In later years, Lord Ranelagh and Lord South, an Irish nobleman, would also be included among the prestigious directors of the company. (Chalmers Letters, *1871–1873 Personals*, p. 672 (December 11, 1873)).

33 Report by Lord Poulett for the First Annual General Meeting of Shareholders of the Imperial Silver Quarries, Ltd., April 20, 1868 (Bancroft Library manuscript collection, Item #pfF 862.3, I42 1868).

34 Chalmers Letters, *1868–1869 Personals*, pp. 157 (September 14, 1868). The promised salary wouldn't materialize, however. Chalmers would later propose to take "£500 *per annum*, paid regularly, in place of £1,000 not paid at all." (*Ibid.*, p. 270, January 24, 1869).

35 *Ibid.*, p. 62 (May 18, 1868) and p. 68 (June 13, 1868).

36 *Ibid.*, p. 30 (April 4, 1868).

37 A cabin passenger list for the *City of Paris*, including Chalmers's name, was published the day after the ship arrived in New York. *New York Times,* September 23, 1867.

38 *New York Times*, October 12, 1867; Chalmers Letters, *1868–1869 Personals*, p. 17 (April 4, 1868).
39 Clemens, S., *The Letters of Mark Twain* (1867–1875) (Vol. 2) (available online through Project Gutenberg, 2006 release, gutenberg.org/files/3194/3194-h/3194-h.htm).
40 *San Francisco Examiner*, November 4, 1867.
41 *Sacramento Daily Union*, November 8, 1867. As luck would have it, Chalmers's path again duplicated that of Sam Clemens with his choice of Sacramento hotels; Clemens, too, had resided at the Orleans Hotel in 1866 while employed at the *Sacramento Union*.
42 Chalmers Letters, 1867–1870 Imperial Officials, p. 391 (August 3, 1869).
43 Nadeau, R., *The Silver Seekers* (Crest Publishers 1999), p. 51.

Chapter 2

1 *Monitor Gazette*, July 30, 1864; *see also* Jonas Winchester's 1872 report on the development of the mines of Alpine County, reproduced in Raymond, R. W., *Silver and Gold: An Account of the Mining and Metallurgical Industry of the United States* (New York, J.B. Ford and Co., 1873) (undated facsimile reproduction, University of Michigan Historical Reprint Series), pp. 92–93.
2 *Monitor Gazette*, September 3, 1864.
3 *Monitor Gazette*, June 11, 1864 (reprinting the 1863 Monitor District Mining Laws); July 7, 1863, Township list (Amador County (CA) Archives, doc. #893.10.83).
4 J. Winchester report, *supra*, p. 93.
5 *Monitor Gazette*, June 11, 1864.
6 *Ibid.*
7 Prospectus for "The "Scottish Chief Silver Mining Company, Limited," January 1868 (Michigan State University Archives and Historical Collections, Jones Collection (c.103, Folder 3)).
8 Letter from O.F. Thornton to Capt. James Jones dated June 10, 1865 (Jones Collection at Michigan State University Archives, *supra* (Folder 2)).
9 S. G. Lewis estimated the population of the entire Monitor Mining District at just 150 in July, 1869. In Raymond, R. W., *Mining Statistics West of the Rocky Mountains* (1869).
10 *Monitor Gazette,* June 25, 1864.
11 *Monitor Gazette*, June 18, 1864; August 20, 1864; and September 17, 1864.
12 *Monitor Gazette*, June 18, 1864.
13 Pierce's Stable and Day & Goff's Hotel are both mentioned in a deed from James F. Pierce to David Stephenson, recorded in Book "B" of Alpine County Records, p. 206 (March 27, 1866).
14 *Alpine Miner*, July 6, 1867.
15 *Alpine Chronicle*, December 28, 1867, and July 29, 1876.
16 Report by Lewis Chalmers, in Raymond, *Mineral Resources West of the Rocky Mountains* (1872); Clark, W. B., *Mines & Mineral Resources of Alpine County, California* (Calif. Div. of Mines and Geology, Sacramento CA, 1977), p. 22.
17 Eno, H., *Twenty Years on the Pacific Slope: Letters of Henry Eno from California and Nevada, 1848–1871* (Yale Univ. Press, New Haven CT, 1965), p. 161 (Letter dated April 12, 1867).
18 *Alpine Chronicle*, April 13, 1867.
19 *Alpine Miner*, May 11, 1867.
20 *Mining and Scientific Press*, June 15, 1867.
21 *Alpine Miner*, April 27, 1867; *Alpine Chronicle*, June 29, 1867.
22 *Mining and Scientific Press*, July 6 and September 7, 1867.
23 This particular excitement at the Tarshish would not last long; by May 1868, the mine had shut down over shareholder disputes. Chalmers Letters, *1867–1870 Imperial Officials*, p. 37 (May 4, 1868). It would reopen in 1869, then periodically close and open again through at least 1879.
24 Chalmers Letters, *1868–1869 Personals*, p. 96 (July 6, 1868). His official correspondence similarly confirmed, "I find there is not a [place of] any description to be got." *1867–1870 Imperial Officials,* p. 65 (May 30, 1868).
25 *1867–1870 Imperial Officials*, pp. 1 and 2 (November 22 and 25, 1867).
26 *1868–1869 Personals*, pp. 118-121 (24 August 1868); *see also 1868–1870 Imperial Officials*, pp. 377–383 (August 30, 1869), discussing the February 23, 1866, Agreement between Jones and Dumas, as agent for the Imperial.
27 Petiss acquired his title on October 15, 1866, by virtue of the sheriff's deed. *1868–1869 Personals*, pp. 120–121 (August 24, 1868).

28 *1868–1869 Personals*, p. 119 (August 24, 1868) and pp. 280–281 (February 1, 1869), indicating that Capt. Jones (as a majority shareholder) had *also* given Imperial Company a purported deed to the Michigan property (which he knew the company did not own) on August 16, 1867.
29 Some of the authorized shares had never been formally issued, but had still been "allotted" on the company's books in the names of third parties. Yet Jones was demanding to be given shares of the Imperial Company in exchange for shares of Michigan standing in those third parties' names—despite providing no formal transfer. *1867–1870 Imperial Officials*, pp. 378–382 (August 30, 1869).
30 *1867–1870 Imperial Officials*, pp. 3–4 (November 27 and December 3, 1867).
31 *Ibid.*, p. 7 (December 9, 1867).
32 Even if got a judgment against him and Frank Jones, "we could not recover." *Ibid.*, p. 102 (July 27, 1868).
33 *1868–1869 Personals*, pp. 17–19 (April 4, 1868).
34 Letter from F.A.S. Jones to James Jones, December 9, 1867. (Jones Collection at Michigan State University Archives, *supra* (Folder 3)).
35 *1868–1869 Personals*, p. 4 (February 29, 1868).
36 Chalmers accused Petiss of being a "tool" of F.A.S. Jones by acquiring "trumped up" salary claims of Jones and Thornton and rolling those into his (Pettis's) judgment. *1868–1869 Personals*, pp. 19–20 (April 4, 1868).
37 *Ibid.*, pp. 18–20 (April 4, 1868); p. 90 (June 29, 1868); pp. 117–118 (August 24, 1868).
38 *1867–1870 Imperial Officials*, p. 90 (June 26, 1868) (confirming Frank Jones had been forced to confess that he'd received the full $5,000); p. 141 (September 21, 1868) (describing Petiss's transfer directly to the Imperial); and p. 216 (February 1, 1869, confirming date Chalmers took possession).
39 Letter from James Jones to the Right Honourable The Earl Poulett (Chairman) and the Directors of the Imperial Silver Quarries Company, Limited, February 4, 1868 (Bancroft Library, #pfF862.3 I43J6).
40 *1867–1870 Imperial Officials*, pp. 238–239 (February 8, 1869); *1868–1869 Personals*, pp. 20–21 (April 4, 1868).
41 Letter from F.A.S. Jones to his brother, James Jones, dated September 29, 1864 (Alpine County Historical Society & Museum, #94.21.02).
42 *1867–1870 Imperial Officials*, p. 36 (April 29, 1868); p. 212 (January 25, 1869); and 238 (February 8, 1869).
43 *Ibid.*, p. 22 (March 27, 1868).
44 *Alpine Chronicle*, February 22, 1868; *1867–1870 Imperial Officials*, p. 216 (February 1, 1869).
45 *1867–1870 Imperial Officials*, p. 39 (May 4, 1868) and p. 131 (September 21, 1868).
46 "Report of First Annual General Meeting of Shareolders of the Imperial Silver Quarries Company, Ltd." by the Rt. Honourable the Earl Poulett, (April 20, 1868) (Bancroft Library, #pfF862.3 I42 1868) (describing the tunnel as 6 feet wide at the bottom tapering to five feet wide at top; and seven feet from bottom of track to centre of the arch, with a 7/8-inch grade in every 16 feet).
47 *1868–1869 Personals*, p. 57 (April 29, 1868).
48 *1867–1870 Imperial Officials*, p. 130 (September 21, 1868); and p. 213 (January 25, 1869).
49 *1868–1869 Personals*, pp. 11–12 (March 25, 1868).
50 *Ibid.*, pp. 40–41 (April 12, 1868).
51 *Ibid.*, p. 51 (April 27, 1868).
52 *1867–1870 Imperial Officials*, p. 35 (May 1, 1868). He similarly wrote to Eliza that the Senate and Fashion were "no go." *1868–1869 Personals*, p. 42 (April 12, 1868).
53 *1867–1870 Imperial Officials*, p. 35 (May 1, 1868).
54 *1868–1869 Personals*, p 61 (May 14, 1868) (estimating the 150-foot contract would be finished "Tuesday or Wednesday"); *1867–1870 Imperial Officials*, p. 54 (May 30, 1868).
55 Letter from O. F. Thornton to W. V. Barkalow, May 25, 1868 (Wickwire private collection, copy in possession of the author).
56 *1867–1870 Imperial Officials*, pp. 54–55 (May 30, 1868) (noting that an additional inducement for avoiding contracting for the work was a recent California act permitting workmen to file mechanics' liens against a project, even if the contractor himself had been paid); p. 97 (July 10, 1868); p. 213, (January 25, 1869) (comparing the Imperial's rate of running to that of other mines in the area).
57 *1867–1870 Imperial Officials*, p. 236 (February 8, 1869).
58 *Ibid.*, pp. 46 (May 8, 1868); 59–60 (May 30, 1868); 241 (February 15, 1869); *see also* Alpine County Recorder's Book "C," pp. 246–247 (purchase of 160 acres from Mary Ready

Hanson on 8 May 1868, recorded 16 February 1869). Title to this property was not firmly established, however. The unused mill site on which Chalmers built his house was later claimed by the original squatter, a man named Oppenheim. *1867–1870 Imperial Officials*, p. 134 (September 21, 1868); p. 234 (February 8, 1869). Chalmers wrote the State Attorney General about perfecting his title, saying he'd had verbal permission to build on the property from a representative of the Mt. Bullion mine, who'd bought the property from Oppenheim, as long as Chalmers left room for the Mt. Bullion's flume-way. (*1867–1870 Imperial Officials*, pp. 166–167 (October 26, 1868). Chalmers later requested the County Recorder certify that "no one was in possession of the ground, the same was not enclosed, and no work done to hold it," presumably to support his title by adverse possession. (*1867–1870 Imperial Officials*, p. 248 (February 20, 1869).

59 *1868–1869 Personals*, p. 39 (April 12, 1868); pp. 71–72 (June 18, 1868); p. 96 (July 6, 1868).

60 Perhaps adding to the site's attractions, a small log cabin already occupied the property. Alpine County Recorder's Book "C," p. 246 (May 8, 1868).

61 *1867–1870 Imperial Officials*, p. 65 (May 30, 1868).

62 *Ibid.*, p. 233 (February 8, 1869).

63 *1868–1869 Personals*, pp. 71–72 (June 18, 1868); p. 92 (June 21, 1868).

64 *1867–1870 Imperial Officials*, p. 73 (June 10, 1868).

65 *1867–1870 Imperial Officials*, p. 220 (February 1, 1869); *see similarly* p. 65 (May 30, 1868).

66 *1868–1869 Personals*, pp. 77–79 (June 19, 1868). Photograph showing a corner of 'Imperial House' was apparently taken by P. W. Greenfield on August 6, 1871, as Chalmers recorded bumping into the photographer while he was at work. *1871–1873 Personals*, p. 213 (August 6, 1871); p. 219 (August 11, 1871).

67 *Ibid.*, pp. 38–39 (April 12, 1868).

68 *Ibid.*, p. 149 (September 5, 1868); p. 367 (May 2, 1869). The property included a dozen trees, though it's unclear whether Chalmers himself had planted them. *1867–1870 Imperial Officials*, p. 167 (October 26, 1868).

69 *1868–1869 Personals*, pp. 149 (September 5, 1868); p. 163 (October 14, 1868).

70 Letter from F. A. S. Jones to James Jones dated July 27, 1868 (Alpine County Historical Society #94.21.07); *1867–1870 Imperial Officials*. p. 106 (August 3, 1868).

71 *1867–1870 Imperial Officials*, p. 117 (August 12, 1868); p. 139 (September 21, 1868).

72 *1868–1869 Personals*, p. 159 (September 14, 1868).

73 *Ibid.*, p. 42 (April 12, 1868); *1867–1870 Imperial Officials*, p. 117 (August 12, 1868).

74 *1868–1869 Personals*, p. 100 (July 15, 1868) (order to R. C. Terry in Sacramento).

75 *Ibid.*, p. 70 (June 13, 1868). A "grass widow" was a woman separated or divorced from her husband; a woman left on her own.

76 *Ibid.*, p. 244 (January 4, 1869); *1867–1870 Imperial Officials*, p. 97 (July 10, 1868); p. 220 (February 1, 1869) (offering to pay the $600 additional cost himself). All told, Chalmers spent $667.07 on his house and office; $457.75 on the assay office and laboratory; and $135.88 on the new blacksmith shop. (*1867–1870 Imperial Officials*, p. 223 (February 1, 1869)).

77 *1867–1870 Imperial Officials*, p. 97 (July 10, 1868).

78 *Ibid.*, p. 352 (July 7, 1869).

79 *Ibid.*, p. 133 (September 21, 1868).

80 Letter from F. A. S. Jones to James Jones dated July 27, 1868 (Alpine County Historical Society #94.21.07); *see also 1867–1870 Imperial Officials*, pp. 60–61 (May 30, 1868) and 97–98 (July 10, 1868).

81 Letters from James Jones to E. J. Coates and Alexander P. McEwen, both dated December 22, 1868 (Michigan State University Collection, *op. cit.*, Folder #4).

82 *1868–1869 Personals*, pp. 115–116 (August 24, 1868); *1867–1870 Imperial Officials*, p. 133 (September 21, 1868).

83 *1867–1870 Imperial Officials*, p. 9 (December 13, 1867); pp. 26, 29 (April 8, 1868); p. 46 (May 11, 1868); p. 59 (May 30, 1868); and p. 86 (June 24, 1868). The equipment was eventually sold for unpaid duties. *1867–1870 Imperial Officials*, pp. 204–205 (January 18, 1869); *and see 1868–1869 Personals*, p. 260 (January 18, 1869).

84 *1867–1870 Imperial Officials*, p. 42–45 (May 7, 1868).

85 *Ibid.*, p. 33 (May 1, 1868) and p. 79 (June 19, 1868).

86 *Ibid.*, p. 30 (April 13, 1868); p. 33 (May 1, 1868); p. 42 (May 7, 1868); p. 119 (August 22, 1868); and p. 139 (September 21, 1868).

87 Chalmers made many references to photographic equipment and his troubles learning how to use it. *See, for example, 1868–1869 Personals*, p. 61 (May 14, 1868); p. 67 (June 13, 1868), p. 72 (June 18, 1868); pp. 88 and 99 (June 29, 1868); p. 129 (August 24, 1868); pp. 175–176

(December 2, 1868); p. 191 (December 21, 1868); p. 244 (January 4, 1869); pp. 283–284 (February 1, 1869), p. 298 (February 26, 1869); and pp. 303–306 (February 28, 1869); *see also 1867–1870 Imperial Officials*, p. 95 (July 2, 1868) (acknowledging that "the Apparatus has arrived.")

88 *1868–1869 Personals*, p. 64 (June 7, 1868). The company may have declined to send a transit as requested, perhaps recommending he find one in the States; in both June and December he wrote letters inquiring about the cost of a new or used transit and chain. *Ibid.*, p. 75 (June 18, 1868) and p. 212 (December 22, 1868).

89 *1867–1870 Imperial Officials*, pp. 134–135 (September 21, 1868).

90 *Ibid.*, p. 66 (May 30, 1868).

91 *Ibid.*, p. 111 (August 8, 1868); p. 134 (September 21, 1868).

92 *Ibid.*, p. 175 (November 13, 1868); p. 181 (December 7, 1868).

93 *1868–1869 Personals*, p. 191 (December 21, 1868); p. 318 (March 15, 1869).

94 *1867–1870 Imperial Officials*, pp. 430–436 (October 7, 1869).

95 *1868–1869 Personals*, p. 172 (22 November 1868).

96 *Ibid.*, pp. 16–17 and 26–27 (April 4, 1868).

97 *1867–1870 Imperial Officials*, pp. 89–92 (June 26, 1868), complaining about Jones and saying he had been "promising men work in the Tunnel."

98 *1868–1869 Personals*, p. 69 (June 13, 1868); pp. 96–97 (July 6, 1868).

99 *Ibid.*, p. 102 (July 15, 1868); p. 339 (April 11, 1869).

100 *Ibid.*, p. 61 (May 14, 1868); p. 97 (July 6, 1868), and *1867–1870 Imperial Officials*, pp. 63–64 (May 30, 1868).

101 Chalmers eventually won the battle over his salary, saying he was "much obliged to Harvey," evidently one of the directors who'd stood up for him. *1868–1869 Personals*, p. 114 (August 24, 1868).

102 *Ibid.*, p. 136 (August 28, 1868); and p. 139 (August 27, 1868).

103 *Ibid.*, pp. 159–160 (September 14, 1868).

104 *Ibid.*, p. 129 (August 24, 1868); *1867–1870 Imperial Officials*, p. 129 (September 14, 1868) and p. 138 (September 21, 1868).

105 *1868–1869 Personals*, p. 160 (September 14, 1868).

106 *Ibid.*, pp. 281–283 (February 1, 1869).

107 By September 1868, a "winding up petition" was being circulated, in hopes of ridding the company of its debenture debt. *1867–1870 Imperial Officials*, p. 137 (September 21, 1868).

108 *Ibid.*, p. 172 (November 3, 1868); p. 176 (November 13, 1868).

109 *1868–1869 Personals*, p. 170 (November 22, 1868).

110 *1867–1870 Imperial Officials*, pp. 185–186 (December 10, 1868).

111 *Ibid.*, p. 195 (December 28, 1868).

112 *Ibid.*, pp. 182–183 (December 7 and 8, 1868); p. 195 (December 21, 1868); and p. 249 (January 10, 1869); *1868–1869 Personals*, p. 243 (January 4, 1869).

113 *1868–1869 Personals*, p. 177 (December 2, 1868).

114 Chalmers would later write to his brother John that the chance to pick up other mining investments for the company had been "the main inducement for my coming here." *Ibid.*, p. 323 (March 27, 1869).

115 *Ibid.*, p. 173 (November 23, 1868).

116 *Ibid.*, p. 130 (August 24, 1868); p. 138 (August 27, 1868). He had expressed interest in the I.X.L. Mine as early as June 1868. (Letters, 1868–69, p. 72–74).

117 *Ibid.*, p. 193 (December 21, 1868).

118 *Ibid.*, pp. 188–189 (December 21, 1868). Chalmers also sent home a flattering report about the nearby Buckeye No. 2 and Accacia lodes: *Ibid.*, pp. 214–234 (December 23, 1868).

Chapter 3

1 *1867–1870 Imperial Officials*, p. 205 (January 18, 1869) (mentioning iron stanchions bedded in molten lead); p. 212 (January 25, 1869); and pp. 248–249, 254 (January 4, 1869); *see similarly* Chalmers's Annual Report in February 1869, *Ibid.*, p. 217-8 (February 1, 1869); and p. 82 (June 22, 1868) (mentioning "iron crampets" to secure 6-inch galvanized iron pipe). Although the tunnel was designed to be wide enough for two tracks, only a single line of rail was apparently actually laid, with the ties merely notched for later installation of a second track. *Ibid.*, p. 217 (February 1, 1869). Men were paid $4 per day for ten-hour shifts, if drilling and blasting were involved, or eight-hour shifts for the heavier work using picks. *Ibid.*, p. 243 (February 15, 1869) and p. 350 (July 7, 1869).

2 *Ibid.*, p. 228 (February 5, 1869).

3 *Ibid.*, p. 197 (January 4, 1869).

4 *1868–1869 Personals*, p. 264 (January 18, 1869).
5 By January 1869, finances were so bad Chalmers was offering to give up a portion of his own salary until the tunnel struck ore. *1867–1870 Imperial Officials*, p. 203 (January 18, 1869); *see similarly 1868–1869 Personals*, p. 252 (January 4, 1869).
6 *1868–1869 Personals*, pp. 249, 253 (January 4, 1869); *see similarly 1867–1870 Imperial Officials*, p. 215 (February 1, 1869).
7 *1868–1869 Personals*, p. 250 (January 4, 1869).
8 *Ibid.*, pp. 270–271 (January 24, 1869); *see also* p. 263 (January 18, 1869) (regarding money troubles at home).
9 *Ibid.*, p. 324 (March 27, 1869).
10 *Ibid.*, p. 321 (March 21, 1869). He defended himself to the company from this charge: *1867–1870 Imperial Officials*, pp. 231–232 (February 8, 1869).
11 *1868–1869 Personals*, pp. 276–278 (February 1, 1869). Chalmers was paying the going rate of $4 per ten-hour day. *1867–1870 Imperial Officials*, p. 350 (July 7, 1869).
12 *1868–1869 Personals*, pp. 271–272 (January 24, 1869).
13 *1867–1870 Imperial Officials*, pp. 249–50, 254 (February 19 and 22, 1869).
14 *Ibid.*, p. 259 (February 21, 1869) (confronting Jones) and p. 284 (March 8, 1869) (about the proposed fee on foreign miners).
15 A printed copy of Chalmers's Annual Report to the Directors dated February 1, 1869 is contained in the Jones Collection at Michigan State University. (Michigan State University Archives and Historical Collections, Jones Collection (c.103, Folder 4).
16 *1868–1869 Personals,* p. 271 (January 24, 1869).
17 *Ibid.*, p. 263 (January 18, 1869); p. 270 (January 24, 1869).
18 *1867–1870 Imperial Officials*, p. 258 (February 21, 1869).
19 *Ibid.*, p. 245 (February 9, 1869).
20 *1868–1869 Personals,* p. 268 (January 18, 1869).
21 *Ibid.*, pp. 298 (February 26, 1869); p. 305 (February 28, 1869).
22 *Ibid.*, pp. 276–277 [*misnumbered in original and located between* pp. 308–309] (March 1, 1869).] *See also* p. 268 (January 18, 1869); p. 274 (January 24, 1869); and pp. 299, 305 (February 28, 1869), in which Chalmers complained about the indoor temperature being below freezing.
23 *1867–1870 Imperial Officials*, pp. 228 and 229, (February 5, 1869).
24 *Ibid.*, p. 287 (March 15, 1869).
25 Chalmers himself had suggested the name Bulliona, after 'Mount Bullion' was declined as having already been taken by a town in Mariposa County, California. *Ibid.*, p. 292 (March 31, 1869). *See also* further mentions of the Bulliona post office, *ibid.* pp. 208–209 (January 22, 1869); p. 317 (May 10, 1869); and p. 336 (June 18, 1869).
26 *Ibid.*, p. 218 (February 1, 1869) and p. 233 (February 8, 1869).
27 *Ibid.*, p. 219 (February 1, 1869); *1868–1869 Personals*, p. 132 (August 24, 1868).
28 *1867–1870 Imperial Officials*, pp. 282–283 (March 8, 1869).
29 *Ibid.*, pp. 401–403 (September 11, 1869).
30 *Ibid.*, p. 279 (March 3, 1869).
31 *Ibid.*, p. 286 (March 15, 1869).
32 *1868–1869 Personals*, p. 325 (March 26, 1869).
33 *Ibid.*, p. 318 (March 15, 1869); p. 324 (March 27, 1869).
34 *Ibid.*, pp. 339–340 (April 11, 1869).
35 *Ibid.*, p. 327 (March 28, 1869).
36 *Ibid.*, pp. 371–373 (May 9, 1869); *see also 1867–1870 Imperial Officials*, p. 310 (May 3, 1869).
37 *1868–1869 Personals*, pp. 365-366 (May 3, 1869) (reproducing Wheeler's letter).
38 *1867–1870 Imperial Officials*, p. 322 (May 17, 1869); p. 335 (June 18, 1869).
39 *Ibid.*, pp. 342 (July 1, 1868); p. 346 (July 3, 1868).
40 *Ibid.*, pp. 438, 439 (October 11, 1869).
41 *Ibid.*, pp. 363, 365 (August 9, 1869).
42 *Ibid.*, p. 388 (September 1, 1869); p. 390 (September 3, 1869).
43 *1869–1871 Personals*, pp. 26, 29 (December 6, 1869).
44 *1867–1870 Imperial Officials*, p. 443 (October 25, 1869); pp. 444–451 (November 1, 1869); p. 524 (December 1, 1869).
45 *Ibid.*, p. 412 (September 28, 1869); pp. 418–420 (October 1, 1869).
46 *Ibid.*, p. 459 (November 8, 1869).
47 *1869–1871 Personals*, p. 17 (November 26, 1869); *1867–1870 Imperial Officials*, p. 460 (November 10, 1869).
48 *1869–1871 Personals*, p. 68 (December 26, 1869). Indeed, Chalmers had supported the concept of reconstruction when he first arrived. *1868–1869 Personals*, p. 17 (April 4, 1868).

49 *1869–1871 Personals*, pp. 40, 42 (December 10, 1869).
50 *Ibid.*, p. 29 (December 6, 1869).
51 *Ibid.*, pp. 53–54 (December 20, 1869); *1867–1870 Imperial Officials*, pp. 530–532 (December 13, 1869).
52 *1869–1871 Personals*, pp. 54–55 (December 20, 1869); *1867–1870 Imperial Officials*, p. 525 (December 1, 1869).
53 *1869–1871 Personals*, p. 44 (December 10, 1869); p. 50 (December 13, 1869).
54 *Ibid.*, p. 71 (December 26, 1869).
55 *Ibid.*, p. 58 (December 18, 1869); p. 66 (December 27, 1869); p. 70 (December 26, 1869).
56 *Ibid.*, p. 70 (December 26, 1869).
57 *Ibid.*, p. 30 (December 6, 1869); p. 35 (December 7, 1869).
58 *Ibid.*, p. 29 (December 6, 1869).
59 *Ibid.*, p. 27 (December 6, 1869).
60 *Ibid.*, pp. 59–60 (December 27, 1869).
61 *Ibid.*, p. 64 (December 27, 1869).
62 *Ibid.*, p. 47 (December 13, 1869).
63 *Ibid.*, p. 73 (December 27, 1869).
64 *Ibid.*, p. 75 (December 27, 1869).
65 *Ibid.*, pp. 64–65 (December 27, 1869).
66 *Ibid.*, p. 72 (December 27, 1869). George H. Ensign was hoping to tap Lake Tahoe to bring water to the west side of the Sierra. He would die in a Stockton insane asylum in 1871.
67 *1867–1870 Imperial Officials*, p. 540 (January 3, 1870) and p. 542 (January 17, 1870).
68 *Ibid.*, p. 542 (January 17, 1870).
69 *1869–1871 Personals*, p. 92 (January 13, 1870).
70 *Ibid.*, p. 129 (January 31, 1870).
71 *Ibid.*, p. 108 (January 17, 1870).
72 *Ibid.*, pp. 114–116 (January 24, 1870).
73 *Ibid.*, p. 137 (February 7, 1870); *1867–1870 Imperial Officials*, p. 544 (January 17, 1870); p. 547 (February 5, 1870). The "stop" order was actually dated January 5, 1870, so it is possible Chalmers deliberately ignored it for several weeks, hoping the order would be countermanded}.
74 *1867–1870 Imperial Officials*, p. 549 (February 10, 1870).
75 *1871–1873 Personals*, p. 21 (April 18, 1871).
76 Report by Lewis Chalmers, reprinted in Raymond, *Mineral Resources West of the Rocky Mountains* (1872).
77 *1867–1870 Imperial Officials*, p. 577 (April 21, 1870), reflecting liens of $1,276 for his miners and $5,675 for himself.
78 *1871–1873 Personals*, p. 329 (November 26, 1871).
79 *1869–1871 Personals*, p. 167 (February 14, 1870). Although Chalmers had requested that his son, Lewie, be sent out with brother John, Lewie would arrive later.
80 *1869–1871 Personals*, pp. 80–81, 84–87 (January 10, 1870).
81 *1869–1871 Personals*, p. 84 (January 10, 1870).
82 *1869–1871 Personals*, pp. 88–89 (January 10, 1870).

Chapter 4

1 *1869–1871 Personals*, p. 290 (March 24, 1870).
2 *Ibid.*, p. 108 (January 17, 1870); p. 167 (February 14, 1870); and p. 170 (February 15, 1870).
3 *Ibid.*, p. 246 (March 9, 1870).
4 *Ibid.*, p. 144 (February 7, 1870).
5 *1867–1870 Imperial Officials*, p. 550 (February 10, 1870).
6 *1869–1871 Personals*, p. 144 (February 7, 1870).
7 *Ibid.*, pp. 172–173 (February 16, 1870).
8 *Ibid.*, p. 176 (February 16, 1870) (*emphasis in original*).
9 *Ibid.*, pp. 202–205 (February 26, 1870) (*emphasis in original*).
10 *Ibid.*, p. 98 (January 17, 1870).
11 *Ibid.*, p. 240 (March 5, 1870) (indicating three tons of ore were being shipped); 249-250 (March 10, 1870) (reflecting three tons of "first" plus half a ton of "medium" grade ore).
12 *Ibid.*, p. 231 (March 5, 1870); p. 244 (March 9, 1870); and 249–250 (March 10, 1870).
13 *Ibid.*, p. 214 (March 3, 1870).
14 *1867–1870 Imperial Officials*, p. 576 (April 21, 1870); Raymond, R. W., *Mineral Resources West of the Rocky Mountains* (U.S. Government Printing Office, 1870), pp. 3–4 (*transcription at Alpine County Museum*).

15 *1869–1871 Personals*, p. 313 (April 9, 1870). Indeed, he reminded Syme, "$50 ore is more than the average of the Comstock." (*Ibid.*, p. 227 (March 5, 1870) (*emphasis in original*).
16 Raymond, *op. cit.*
17 *1869–1871 Personals*, p. 219 (March 1, 1870).
18 *Ibid.*, p. 222 (March 5, 1870).
19 *Ibid.*, pp. 217–218 (March 1, 1870) (report by John Chalmers for period January 3 to February 28, 1870).
20 *Ibid.*, p. 146 (February 7, 1870).
21 *Ibid.*, p. 167 (February 14, 1870).
22 *Ibid.*, p. 234 (March 5, 1870).
23 *Ibid.*, p. 228 (March 5, 1870).
24 *Ibid.*, pp. 197–199; 209 (February 28, 1870); p. 228 (March 5, 1870).
25 *Ibid.*, p. 236 (March 5, 1870) (*emphasis in original*).
26 *Ibid.*, pp. 207–208 (February 28, 1870); pp. 180–181 (February 18, 1870).
27 *Ibid.*, p. 306 (March 30, 1870); p. 197 (February 28, 1870).
28 *Ibid.*, p. 197 (February 28, 1870); p. 131 (February 2, 1870); and p. 140 (February 7, 1870).
29 *1867–1870 Imperial Officials*, pp. 555–558, 562 (April 18, 1870).
30 *Ibid.*, pp. 555–558.
31 *1869–1871 Personals*, p. 245 (March 9, 1870).
32 *Ibid.*, p. 180 (February 18, 1870).
33 *Alpine Chronicle,* November 5, 1870 (notice of Sheriff's sale in action by Chalmers against The Imperial Silver Quarries Co., Ltd., for $5,951.81 plus costs); *see also 1867–1870 Imperial Officials*, p. 551 (February 5, 1870) (an accounting showing $5,675.46 owed to Chalmers as of the date work on the Imperial was suspended).
34 *1871–1873 Personals*, pp. 12, 26 (April 18, 1871).
35 *Ibid.*, p. 9 (April 18, 1871).
36 *Ibid.*, p. 18 (April 18, 1871) (*emphasis in original*); pp. 90–91 (May 21, 1871); p. 96 (May 28, 1871); p. 167 (July 9, 1871); *see also 1870–1873 Exchequer Officials*, p. 79 (June 1, 1870) (mentioning 'stories afloat' about him back in London.
37 *1871–1873 Personals*, pp. 121–122 (June 10, 1871).
38 *Ibid.*, p. 167 (July 9, 1871).
39 *Ibid.*, pp. 9–10 (April 18, 1871).
40 *Ibid.*, p. 76; 79 (May 19, 1871) (*emphasis in original*).
41 *Ibid.*, p. 80 (May 19, 1871).
42 *Ibid.*, p. 128 (June 12, 1871).
43 *Ibid.*, p. 86 (May 12, 1871).
44 *Ibid.*, p. 98 (May 28, 1871).
45 *Ibid.*, p. 102 (May 28, 1871).
46 *Ibid.*, p. 111 (May 29, 1871).
47 *Ibid.*, p. 77 (May 19, 1871).
48 *Ibid.*, p. 86 (May 21, 1871).
49 *Ibid.*, pp. 87, 89 (May 21, 1871). Eliza had borrowed money from A. P. MacEwen to help Chalmers pay his debts before he left London, possibly as a straw borrower. It is unclear whether that debt figured into her financial distress. *See also Ibid.*, pp. 17–18 (April 18, 1871).
50 *Ibid.*, p. 95 (May 28, 1871).
51 *Ibid.* p. 296 (September 30, 1871).
52 *Ibid.*, pp. 103–110 (May 29, 1871); p. 102 (May 28, 1871).
53 *Ibid.*, p. 144 (June 22, 1871) (*emphasis in original*).
54 *Ibid.*, p. 111 (May 29, 1871).
55 *Ibid.*, p. 75 (May 19, 1871).
56 *Ibid.*, p. 62 (May 22, 1871).
57 *Ibid.*, p. 87 (May 21, 1871).
58 *Ibid.*, p. 62 (May 22, 1871); p. 66 (May 17, 1871).
59 *Ibid.*, p. 83 (May 19, 1871).
60 *Ibid.*, p. 135 (June 18, 1871); p. 165 (July 9, 1871).
61 Lewis and John Chalmers likely made the move from Bulliona to Silver Creek about May 1871. Lewie's exact arrival date is unclear, but a brief reference in a letter to Lewie confirms he was residing with Chalmers by mid-May 1871. *Ibid.*, p. 82 (May 21, 1871).
62 *Ibid.*, p. 263 (September 3, 1871).
63 *Ibid.*, pp. 207, 211 (August 6, 1871); *see also Ibid.*, pp. 178–182 (July 24, 1871).
64 *Ibid.*, p. 181 (July 24, 1871) (*emphasis in original*).
65 *Ibid.*, p. 173 (July 16, 1871) (*emphasis in original*).
66 *Ibid.*, p. 178 (July 24, 1871).

67 *Ibid.*, p. 195 (July 27, 1871); p. 295 (September 30, 1871).
68 *Ibid.*, p. 203 (August 1, 1871); p. 217 (August 11, 1871).
69 *Ibid.*, p. 230 (August 17, 1871).
70 *Ibid.*, p. 228 (August 17, 1871); p. 263 (September 3, 1871) (*emphasis in original*).
71 *Ibid.*, p. 241 (August 21, 1871); p. 255 (August 30, 1871) (indicating Chalmers had just returned from Mono County the previous day); p. 256 (August 30, 1871); p. 258 (September 1, 1871); and pp. 271–273 (September 10, 1871).
72 *1871–1873 Personals*, p. 272 (September 11, 1871); p. 287 (September 17, 1871).
73 *Ibid.*, p. 259 (September 3, 1871).
74 *Ibid.*, p. 293 (September 30, 1871) (*emphasis in original*); *see also* p. 197 (July 30, 1871) (regarding mudslide damage to the flume after a "violent thunder storm").
75 *Ibid.*, p. 273 (September 13, 1871).
76 *Ibid.*, p. 295 (September 30, 1871).
77 *Ibid.*, p. 333 (November 26, 1871).
78 *1870–1873 Exchequer Officials*, p. 508 (December 15, 1871); *and see 1871–1873 Personals*, p. 318 (November 15, 1871); p. 319 (November 21, 1871); p. 335 (November 26, 1871); p. 339 (December 15, 1871).
79 Letter from George Lowe to Oscar F. Thornton (December 31, 1871), Judy Wickwire collection (copy in possession of the author).
80 The I.X.L. was claimed on July 1, 1861, the same day as the Buckeye No. 1 was taken up. †*Ibid.*, pp. 48–49, and pp. 54, 56 (May 22, 1871); and p. 67 (May 17, 1871)†. An 1872 report described the I.X.L. as "a strong lode, beautifully cased with clay selvages between highly polished walls," hopefully predicting that "another year will find both [I.X.L. and Exchequer] on the dividend-paying list." Raymond, Rossiter W., *Mineral Resources West of the Rocky Mountains* (U.S. Government Printing Office, 1872), p. 2 (*transcription at Alpine County Museum*).
81 *1871–1873 Personals*, pp. 352–356 (January 14, 1872); p. 358 (January 13, 1872); p. 373 (January 17, 1872).
82 *1871–1873 Personals*, pp. 360, 366 (January 15, 1872).
83 *Ibid.*, p. 371 (January 16, 1872); p. 373 (January 17, 1872).
84 *Ibid.*, p. 383 (February 26, 1872); p. 385 (March 5, 1872); p. 388 (March 26, 1872).
85 *Ibid.*, p. 397 (May 22, 1872).
86 Letter by George Lowe to Oscar Thornton (May 5, 1872) (Judy Wickwire collection, copy in possession of the author).
87 Letter from Mrs. Mary Hanson to Oscar F. Thornton (June 17, 1872) (Alpine County Historical Society, Thornton collection).
88 *1871–1873 Personals*, p. 402 (June 17, 1872); p. 403 (June 19, 1872); p. 405 (June 19, 1872).
89 *Ibid.*, p. 409 (June 21, 1872) (relocating the Buckeye No. 1 claim); p. 411 (undated letter approximately June 27, 1872); and p. 438 (September 18, 1872) (forwarding title documents).
90 *Ibid.*, p. 433 (August 24, 1872).
91 *Ibid.*, p. 430 (August 22, 1872); and p. 431 (August 24, 1872).
92 *Ibid.*, pp. 503–504 (February 12, 1873); p. 525 (March 17, 1873) (*emphasis in original*).
93 *Ibid.*, p. 518 (March 17, 1873).

Chapter 5

1 *1870–1873 Exchequer Officials*, p. 617 (October 14, 1872); p. 627 (November 4, 1872); p. 632 (November 15, 1872).
2 *1871–1873 Personals*, pp. 465, 470 (November 8, 1872); p. 493 (November 29, 1872); p. 497 (December 16, 1872).
3 *1870–1873 Exchequer Officials*, pp. 679–681 (January 1, 1873). Regarding the Freyberg system originally used by the mill, *see 1871–1873 Personals*, p. 531 (March 21, 1873).
4 *1870–1873 Exchequer Officials*, p. 681 (January 1, 1873).
5 *Ibid.*, p. 682.
6 *Ibid.*, p. 685.
7 *1871–1873 Personals*, p. 569 (May 15, 1873).
8 *Ibid.*, pp. 466, 471, 473–474 (November 8, 1872).
9 *1871–1873 Personals*, p. 495 (December 9, 1872); *see also* pp. 472–475 (November 8, 1872); p. 564 (April 9, 1873); p. 569 (May 15, 1873).
10 *Ibid.*, p. 493 (November 29, 1872); p. 501 (February 10, 1873); pp. 506–510 (February 12, 1873); *see also 1870–1873 Exchequer Officials*, pp. 659–666 (January 13, 1873); pp. 667–669 (February 2, 1873).

11 *1871–1873 Personals*, p. 516 (February 28, 1873); *see similarly* pp. 511, 513–514 (February 17, 1873); p. 559 (March 23, 1873).
12 *Ibid.*, p. 507 (February 12, 1873).
13 *Ibid.*, p. 502 (February 12, 1873).
14 *Ibid.*, p. 522 (March 17, 1873).
15 *Ibid.*, p. 503 (February 12, 1873) (*emphasis in original*).
16 *Ibid.*, pp. 506, 509 (February 12, 1873).
17 *Ibid.*, p. 517 (March 17, 1873).
18 *Ibid.*, p. 526 (March 17, 1873). Chalmers would not be without a housekeeper for long. By early June, "an old widow lady" was arriving "to-night." *Ibid.*, p. 576 (June 6, 1873).
19 *Ibid.*, pp. 527-539 (March 21, 1873).
20 *Ibid.*, p. 539.
21 *Ibid.*, p. 557.
22 *1873–1874 Exchequer Officials* [Bancroft Library microfilm], p. 1 (March 3, 1873); p. 11 (March 17, 1873).
23 *Ibid.*, p. 12 (March 24, 1873).
24 *Ibid.*, p. 12 (March 24, 1873).
25 *1871–1873 Personals*, p. 522 (March 17, 1873).
26 *Ibid.*, p. 524.
27 *Ibid.*, p. 525.
28 *Ibid..*, pp. 559–561 (March 23, 1873).
29 *1873–1874 Exchequer Officials* [Bancroft Library microfilm], p. 70 (June 30, 1873); pp. 126–129 (September 15, 1873).
30 *Ibid.*, pp. 126–127 (September 15, 1873); *1871–1873 Personals*, p. 579 (June 13, 1873).
31 *1871–1873 Personals*, pp. 573–574 (May 15, 1873); pp. 578–579 (June 13, 1873).
32 *Ibid.*, p. 580 (June 30, 1873).
33 *Ibid.*, p. 582 (July 14, 1873). 'Mullers' are horizontal revolving arms.
34 *Ibid.*, p. 592 (July 23, 1873).
35 *Ibid.*, pp. 589–590 (July 23, 1873).
36 *Ibid.*, pp. 590–591 (July 23, 1873). Both this photograph and the previous one were taken by A. O. Friend of Roseville, California. *Ibid.*, p. 655 (December 1, 1873).
37 Raymond, R. W., *Mineral Resources West of the Rocky Mountains* (U.S. Government Printing Office, 1873), p. 5 (*transcription at Alpine County Museum*); *1873–1874 Exchequer Officials* [Bancroft Library microfilm], p. 130 (September 15, 1873).
38 *Ibid.*, pp. 130–132 (September 15, 1873).
39 *1871–1873 Personals*, p. 596 (September 9, 1873) (*emphasis in original*).
40 *Ibid.* pp. 601–602 (September 12, 1873).
41 *Ibid.*, p. 608 (October 30, 1873). This same letter reflected a *different* boiler failure at the I.X.L. Mine, when the tubers burst in a small, upright boiler, caused by a "rush of water" that pressed the machine to extremes. A replacement boiler had just been installed as of the date of that letter, however, and he hoped to unwater the mine by the following week.
42 *1873–1874 Exchequer Officials* [Bancroft Library microfilm], p. 131 (September 15, 1873).
43 *1871–1873 Personals*, pp. 598–599 (September 10, 1873).
44 *1873–1874 Exchequer Officials* [Bancroft Library microfilm], p. 143 (September 30, 1873).
45 Chalmers variously describes the value of the silver bar as $464.18 and $442.37. *1873–1874 Exchequer Officials* [Bancroft Library microfilm], pp. 140–141 (September 22, 1873); p. 118 (September 15, 1873); p. 148 (September 30, 1873).
46 *Ibid.* p. 143 (September 30, 1873).
47 *Ibid.*, pp. 149–150 (September 30, 1873) (*emphasis in original*); *see also* p. 143 (September 30, 1873) (protesting "the injustice" of the charges).
48 *Ibid.*, pp. 156 (October 17, 1873); p. 165 (October 27, 1873).
49 *Ibid.*, p. 160 (October 27, 1873).
50 *Ibid.*, p. 164.
51 *Ibid.*, p. 174 (October 29, 1873).
52 *1871–1873 Personals*, p. 613–616 (November 2, 1873) (estimating the cost at $20,782); p. 622 (November 10, 1873); p. 630 (November 15, 1873); p. 634 (November 17, 1873). *See similarly 1873–1874 Exchequer Officials* [Bancroft Library microfilm], pp. 164–165 (October 27, 1873); p. 200 (November 17, 1873).
53 *1871–1873 Personals*, p. 612 (November 2, 1873) (inaccurately mentioning £3,000); and p. 619 (November 2, 1873) (confirming the actual offer had been $3,000).
54 *1871–1873 Personals*, p. 620 (November 3, 1873).
55 *Ibid.*, p. 628 (November 12, 1873); p. 640 (November 23, 1873).
56 *1873–1874 Exchequer Officials* [Bancroft Library microfilm], p. 191 (November 12, 1873).

57 *Ibid.*, p. 196 (November 17, 1873).
58 *1871–1873 Personals*, p. 627 (November 12, 1873).
59 *Ibid.* p. 617 (November 2, 1873).

Chapter 6

1 *1873–1874 Exchequer Officials* [Bancroft Library microfilm], p. 201 (November 24, 1873).
2 *Ibid.*
3 *Alpine Chronicle*, January 3, 1874; April 4, 1874.
4 *1873–1874 Exchequer Officials* [Bancroft Library microfilm], p. 201 (November 24, 1873); p. 206 (December 1, 1873).
5 *Ibid.*, p. 202 (November 24, 1873).
6 *1871–1873 Personals*, p. 655 (December 1, 1873).
7 *1873–1874 Exchequer Officials* [Bancroft Library microfilm], p. 216 (December 1, 1873).
8 *Ibid.*, p. 222 (December 3, 1873); p. 248 (December 15, 1873).
9 *Ibid.*, pp. 248–249 (December 15, 1873).
10 *Ibid.*, p. 233 (December 7, 1873).
11 *Ibid.*, p. 221 (December 3, 1873) (*emphasis in original*).
12 *Ibid.*, p. 222.
13 *Ibid.*, p. 233 (December 7, 1873).
14 *1871–1873 Personals*, p. 664 (December 3, 1873); p. 670 (December 8, 1873).
15 *Ibid.*, pp. 656–659 (December 1, 1873).
16 *1873–1874 Exchequer Officials* [Bancroft Library microfilm], pp. 196–197 (November 17, 1873); pp. 208-210 (December 2, 1873).
17 *Ibid.*, p. 211 (December 2, 1873).
18 *Ibid.*, p. 242 (December 10, 1873).
19 *Ibid.*, pp. 237–241 (December 10, 1873).
20 *Ibid.*, p. 248 (December 15, 1873) (*emphasis in original*).
21 *Ibid.*, p. 256 (December 14, 1873).
22 *1871–1873 Personals*, pp. 677–678 (December 17, 1873).
23 *Ibid.*, p. 689 (December 28, 1873) (*emphasis in original*).
24 *Ibid.*, p. 694 (December 29, 1873).
25 *1873–1874 Exchequer Officials* [Bancroft Library microfilm], p. 267 (December 25, 1873).
26 *Ibid.*, p. 269 (December 29, 1873). That, at least, was welcome news.
27 *Ibid.*, p. 176 (October 30, 1873).
28 *Ibid.*, pp. 203–204 (November 24, 1873).
29 Casson, M. and Spence, C. C., *British Investments and the American Mining Frontier, 1860–1901* (2000, Taylor & Francis Co.), p. 199 (*Google Books*).
30 *1873–1874 Exchequer Officials* [Bancroft Library microfilm], p. 207 (December 1, 1873).
31 *1871–1873 Personals*, pp. 674–675 (December 11, 1873).
32 *Ibid.* p. 690 (December 28, 1873).
33 *1873–1874 Exchequer Officials* [Bancroft Library microfilm], p. 253 (December 17, 1873).
34 *1871–1873 Personals*, pp. 690–692 (December 28, 1873); pp. 693–694 (December 29, 1873); *1873–1874 Exchequer Officials* [Bancroft Library microfilm], p. 248 (December 15, 1873); p. 261 (December 22, 1873); p. 268 (December 25, 1873).
35 *1873–1874 Exchequer Officials* [Bancroft Library microfilm], p. 261 (December 22, 1873).
36 *Ibid.*, pp. 251–253 (December 17, 1873).
37 *Ibid.*, p. 252 (December 17, 1873).
38 *Ibid.*, pp. 261–262 (December 22, 1873).
39 *Ibid.*, p. 265 (December 25, 1873).
40 *1871–1873 Personals*, p. 684 (December 22, 1873). *See also 1873–1874 Exchequer Officials* [Bancroft Library microfilm], p. 261 (December 22, 1873). Perhaps to save future rebellions, Chalmers would later hire contractors to clear the road. *Alpine Chronicle*, February 28, 1874.
41 Letter from Samuel A. Chapin to Oscar Thornton, December 23, 1873 (Wickwire private collection, copy in possession of the author).
42 *1871–1873 Personals*, pp. 678–9 (December 18, 1873).
43 *1873–1874 Exchequer Officials* [Bancroft Library microfilm], p. 272 (December 30, 1873); p. 276 (January 3, 1874).
44 *Alpine Chronicle*, February 14, 1874.
45 *Alpine Chronicle*, March 7, 1874; *see also* March 28, 1874.
46 *Alpine Chronicle*, February 21, 1874.

47 Chalmers had paid Adams McNeil for goods delivered to the Exchequer Company, but had not paid for the I.X.L. supplies. *1873–1874 Exchequer Officials* [Bancroft Library microfilm], p. 447 (May 26, 1874).
48 *Alpine Chronicle*, May 16, 1874.
49 *Alpine Chronicle*, April 4, 1874.
50 *Alpine Chronicle*, April 11, 1874.
51 *Alpine Chronicle*, May 30, 1874.
52 *Alpine Chronicle*, May 30, 1874.
53 *Alpine Chronicle*, June 20, 1874.
54 *Alpine Chronicle*, June 20, 1874; June 27, 1874.
55 Raymond, R. W., *Mineral Resources West of the Rocky Mountains* (U.S. Government Printing Office, 1874), p. 12 (transcription at Alpine County Museum).
56 Raymond, R. W., *Report on the Exchequer Gold & Silver Mining Co. Limited* (New York, September 16, 1874), *copy at* Bancroft Library.
57 *1873–1874 Exchequer Officials* [Bancroft Library microfilm], p. 560 (September 10, 1874).
58 *1873–1874 Exchequer Officials* [Bancroft Library microfilm], pp. 560–565 (September 10, 1874) (*emphasis in original*).
59 *Ibid.*, pp. 560–561.
60 *Ibid.*, pp. 560–563.
61 *Alpine Chronicle*, January 3, 1874.
62 *1873–1874 Exchequer Officials* [Bancroft Library microfilm], p. 565 (September 10, 1874) (*emphasis added*).

Chapter 7

1 Jackson, W. T., "Report on the History of the Grover Hot Springs State Park Area and Surrounding Region of Alpine County" (California Department of Parks and Recreation, Division of Beaches and Parks, 1964) (typescript), pp. 71, 75).
2 Letter from W.G. Romaine to William C. Ralston, 25 June 1875 (Bancroft Library, Willian Chapman Ralston Collection, MSS # 77/88c). If Romaine indeed stopped to see Ralston in San Francisco, he was one of the last to do so. Ralston would commit suicide on August 27 that same year, following the collapse of his financial empire.
3 *1880–1881 Personals*, p. 514 (November 3, 1880).
4 Dustman, K., *Silver Mountain City: Ghost of the Sierra* (Clairitage Press 2011), p. 136 & ftn. 31 and 32.
5 The gold/silver ratio had risen to $31.60 by 1895, per MeasuringWorth.com, tinyurl.com/atcjj8ja.
6 Alpine County Supervisors Minutes, Book "B," pp. 8–11 (October 25, 1875).
7 Chalmers would later write that Antoinette "came here a year ago," suggesting she'd arrived in the summer of 1875. But given that her husband had died that August, it's more likely that she came to Alpine in mid-to-late Fall 1875. *1876–1878 Personals*, pp. 14–15 (July 21, 1876) (transcription by Richard Race, copy at Alpine County Historical Society).
8 *Ibid.*, p. 2 (May 8, 1876).
9 *Ibid.*, p. 9 (June 8, 1876).
10 *Ibid.*, p. 1 (May 12, 1876); p. 9 (June 9, 1876).
11 *Ibid.*, p. 8 (June 6, 1876).
12 *Ibid.*, p. 5 (May 31, 1876).
13 *Ibid.*, p. 3 (May 22, 1876).
14 *Ibid.*, p. 7 (June 6, 1876) (*emphasis in original*).
15 *Alpine Chronicle*, July 29, 1876.
16 *1876–1878 Personals,* pp. 5–6 (May 31, 1876); p. 8 (June 6, 1876).
17 *Ibid.*, pp. 10–11 (June 26, 1876).
18 *Ibid.*, pp. 27–28 (November 3, 1876).
19 *Ibid.*, pp. 14–15 (July 21, 1876).
20 *Ibid.*, p. 16 (July 28, 1876).
21 *Ibid.*, pp. 22–23 (September 6, 1876).
22 *Chronicle*, July 8; July 15; and July 29, 1876).
23 *Alpine Chronicle*, September 2, 1876.
24 *1876–1878 Personals,* p. 30 (December 4, 1876); *see also Alpine Chronicle*, August 26, 1876; September 2, 1876).
25 *Alpine Chronicle*, July 22, 1876.
26 *Alpine Chronicle*, July 22, 1876; September 23, 1876; September 30, 1876; December 9, 1876.

27 This boarding house was likely at the old Nye ranch. *Alpine Chronicle*, September 2, 1876.
28 *Alpine Chronicle*, July 1; July 22; July 29; and August 12 1876.
29 *Alpine Chronicle*, July 15, 1876.
30 *Alpine Chronicle*, July 29, 1876.
31 Letter from Lewis Chalmers to Board of Equalization, July 21, 1876 (original at Alpine County Historical Society).
32 *Alpine Chronicle*, August 5 and 12, 1876.
33 *1876–1878 Personals*, p. 15 (July 21, 1876).
34 *Ibid.*, p. 24 (September 6, 1876).
35 *Ibid.*, p. 35 (undated letter, *circa* March 1, 1877).
36 *Alpine Chronicle*, July 29, 1876.
37 *Alpine Chronicle*, August 12 and 26, 1876; September 2 and 23, 1876.
38 *1876–1878 Personals,* p. 25 (October 18, 1876); *Alpine Chronicle*, October 14, 1876; *Carson Valley News*, October 13, 1876.
39 *1876–1878 Personals*, p. 29 (November 3, 1876).
40 *Ibid.*, p. 29 (November 10, 1876).
41 *Alpine Chronicle*, September 16, 1876.
42 *Alpine Chronicle*, September 9, 1876.
43 *Alpine Chronicle*, October 21, 1876.
44 Gaertner, E. G., "Report on the Advance Mine" (San Francisco, September 1876) (Bancroft Library manuscript collection), pp. 3, 5.
45 *Alpine Chronicle*, September 9 1876.
46 *Alpine Chronicle*, September 30, 1876; *see similar* predictions of a V&T extension: *Alpine Chronicle*, December 23, 1876.
47 *Alpine Chronicle*, October 14, 1876.
48 *Alpine Chronicle*, October 14, 1876; November 4, 1876.
49 *1876–1878 Personals,* p. 26 (October 15, 1876).
50 *Ibid.*, p. 31 (November 29, 1876).
51 *Alpine Chronicle*, October 21, 1876.
52 *1876–1878 Personals*, p. 31 (December 15, 1876).
53 *Ibid.*, p. 32 (February 21, 1877).
54 *Alpine Chronicle*, October 28, 1876; *1873–1874 Exchequer Officials* [Bancroft Library microfilm], p. 127 (*circa* September 15, 1873).
55 Jackson, W. T., "Report on the History of the Grover Hot Springs State Park Area and Surrounding Region of Alpine County" (California Department of Parks and Recreation, Division of Beaches and Parks, 1964) (typescript), p. 72.
56 *1876–1878 Personals*, pp. 25–26 (October 15, 1876).
57 *Alpine Chronicle*, December 16, 1876. Most of these eleven lodes were part of the old Imperial's set of claims. But at least one was a claim Chalmers himself had located in July 1875.
58 *Ibid.*
59 *Alpine Chronicle,* December 30, 1876.
60 *Ibid.*
61 *Alpine Chronicle*, January 6, 1877.
62 *1876–1878 Personals*, p. 32 (February 21, 1877).
63 *Ibid.*, p. 37 (March 23, 1877); *Carson Valley News*, March 20, 1877.
64 *1876–1878 Personals*, p. 34 (February 21, 1877).
65 *Ibid.*, p. 33 (February 21, 1877).
66 *Ibid.*, p. 35 (*circa* March 1, 1877).
67 *Ibid.*, pp. 37–38 (April 11, 1877).
68 Sheriff's Deed dated June 23, 1879, reflecting Judgment and Writ of Execution in favor of Arnot. California State Library, Syme Collection (SMC Box 21, Folder 1) (copy in possession of the author).
69 *1876–1878 Personals*, p. 40 (May 20, 1877); *see also Alpine Chronicle*, December 15, 1877.
70 *Alpine Chronicle*, December 15, 1877.
71 Letter from R. K. Love to Thomas Hay, May 9, 1877 (Dennis Nelson collection, copy in possession of the author).
72 *1876–1878 Personals*, p. 40 (May 20, 1877).
73 *Ibid.*, p. 47 (June 27, 1877).
74 *Ibid.*, p. 43 (June 18, 1877).
75 *Ibid.*, p. 41 (June 12, 1877).
76 *Ibid.*, p. 42 (June 17, 1877).
77 *Ibid.*, p. 43 (June 18, 1877).

78 *Ibid.*, p. 44–46 (June 22, 1877) (*emphasis in original*).
79 *Ibid.*, p. 47 (June 27, 1877).
80 *Ibid.*, p. 60 (August 25, 1877).
81 *Ibid.*, p. 51 (July 8, 1877).
82 *Ibid.*, p. 52 (July 23, 1877).
83 *Alpine Chronicle*, December 15, 1877.
84 *1880–1881 Personals*, p. 230 (July 12, 1880), explaining that the Arnots had foreclosed (acquiring title in the name of N. D. Arnot, Sr.'s wife), then reconveyed it through a friendly third party to trustees for Exchequer Company debenture holders.
85 *1876–1878 Personals*, p. 54 (July 26, 1877).
86 *Ibid.*, p. 59 (August 15, 1877).
87 *Ibid.*, p. 60 (August 25, 1877); p. 61 (August 24, 1877).
88 *Ibid.*, pp. 62–63 (August 24, 1877).
89 *Ibid.*, p. 62.
90 *Ibid.*
91 *Ibid.*, p. 50 (July 8, 1877).
92 *Ibid.*, p. 63 (August 25, 1877).
93 *Ibid.*, pp. 66–67 (September 4, 1877).
94 *Ibid.*, p. 68 (September 9, 1877).
95 *Ibid.*, p. 69 (September 10).
96 *Ibid.*, p. 71 (September 18, 1877); p. 75 (October 12, 1877).
97 *Ibid.*, pp. 75–76 (October 12, 1877).
98 *Ibid.*, p. 81 (October 19, 1877).
99 *Ibid.*, p. 82 (October 22, 1877).
100 *Ibid.*, p. 84 (October 29, 1877); p. 86 (November 7, 1877); p. 88 (November 15, 1877).
101 *Ibid.*, pp. 88–89 (November 15, 1877).
102 *Ibid.*, p. 91 (November 22, 1877).
103 *Ibid.*, p. 92 (November 26, 1877).
104 Arnot family history (manuscript), at Alpine County Historical Society.
105 *Alpine Chronicle*, December 15, 1877.
106 *1876–1878 Personals*, p. 98 (December 19, 1877).
107 *Ibid.*, p. 99.
108 *1878–1880 Personals*, p. 70 (July 27, 1878).
109 *Alpine Chronicle*, February 9, 1878; *1878 Personals*, p. 3 (January 28, 1878).
110 *1878 Personals*, p. 2 (January 28, 1878).
111 *1881–1884 Personals*, p. 203 (December 16, 1881) (confirming Chalmers was in London from December 30, 1877 to June 1, 1878).

Chapter 8

1 *1878 Personals (Vol. 1)*, p. 15 (February 4, 1878).
2 *Ibid.*, pp. 1–3 (January 28, 1878).
3 *Ibid.*, p. 46 (February 16, 1878).
4 *Ibid.*, p. 3 (January 28, 1878).
5 *Ibid.*, p. 4 (January 28, 1878).
6 *Ibid.*, p. 32 (February 8, 1878).
7 *Ibid.*, pp. 3–4 (January 28, 1878).
8 *Ibid.*, p. 44 (February 16, 1878).
9 *Ibid.*, p. 50 (February 20, 1878).
10 *Ibid.*, p. 61 (February 27, 1878).
11 *Ibid.*, p. 82 (March 9, 1878).
12 *Ibid.*, p. 19 (February 5, 1878) (*emphasis in original*).
13 *Alpine Chronicle*, February 16, 1878.
14 *Alpine Chronicle*, March 23, 1878.
15 *Alpine Chronicle*, February 2, 1878; February 16, 1878; April 13, 1878.
16 *1878 Personals* (Vol. 1), p. 13 (February 4, 1878); p. 137 (March 26, 1878) (indicating Chalmers had borrowed money from Arnot to get to San Francisco).
17 *Ibid.*, pp. 13–14 (February 4, 1878); p. 93 (March 15, 1878).
18 *Ibid.*, p. 14 (February 4, 1878).
19 *Ibid.*, p. 40 (February 13, 1878).
20 *Ibid.*, p. 21 (February 5, 1878).
21 *Ibid.*, p. 34 (February 11, 1878); *see also* p. 42 (February 15, 1878).
22 *Ibid.*, p. 42 (February 15, 1878).

23 *Ibid.*, pp. 243–244 (undated letter *circa* May 13, 1878).
24 *Ibid.*, p. 113 (March 19, 1878).
25 *Ibid.*, p. 28 (February 8, 1878). A similar debenture deal was likely offered to Exchequer investors.
26 *Ibid.*, p. 68 (March 2, 1878); p. 202 (April 9, 1878).
27 *Ibid.*, p. 87 (March 12, 1878); pp. 104–105 (March 18, 1878).
28 *Ibid.*, p. 104 (March 18, 1878).
29 *Ibid.*, p. 105 (March 18, 1878).
30 *Ibid.*, p. 69 (March 2, 1878).
31 Poulett held 2,543 shares of Exchequer. *Ibid.*, pp. 74–75 (March 5, 1878).
32 *Ibid.*, p. 91 (March 14); p. 95 (March 15, 1878).
33 *Ibid.*, p. 146 (March 29, 1878).
34 *Ibid.*, p. 149 (March 30, 1878).
35 *Ibid.*, p. 150–151 (March 30, 1878) (*emphasis in original*).
36 *Ibid.*, p. 165 (April 4, 1878); *see also* p. 164 (April 1, 1878).
37 *Ibid.*, pp. 167–168 (April 5, 1878).
38 *Ibid.*, pp. 174–175 (April 5, 1878))
39 *1878–1880 Personals*, pp. 21–22 (May 17, 1879) (transcription by Richard Race, copy at Alpine County Historical Society); *1878 Personals (Vol. 2)*, p. 11 (June 13, 1878); pp. 36–37 (July 3, 1878).
40 *Ibid.*
41 *1878 Personals (Vol. 1)*, p. 223 (April 19, 1878). John J. and Isabella Kelly lived at 2512 Mission Street, San Francisco. *1880–1881 Personals, p. 233* (July 13, 1880). The Kellys were a respectable couple with two young daughters. The 1880 Census for San Francisco described Kelly's occupation as a 'machinist' (Enumeration District 177, p. 23, tinyurl.com/3y5r5s99). The 1870 Census similarly described him as an 'engine builder.' (2nd Precinct, 10th Ward, tinyurl.com/pkdzt27y).
42 *Ibid.*, p. 232 (April 23, 1878).
43 *Ibid.*, p. 234 (April 24, 1878).
44 *Carson Valley News*, May 31, 1878; June 7, 1878; *see similarly Sacramento Daily Union*, June 6, 1878.
45 *1878 Personals (Vol. 1)*, p. 124 (March 21, 1878); p. 140 (March 26, 1878).
46 *Ibid.*, pp. 174–175 (April 5, 1878); p. 212 (April 15, 1878).
47 *Ibid.*, p. 172 (April 5, 1878).
48 *Ibid.*, p. 230 (April 23, 1878).
49 *Ibid.*, p. 136–137 (March 26, 1878).
50 *Ibid.*, pp. 212–213 (April 15, 1878).
51 *Ibid.*, p. 224 (April 19, 1878).
52 *Ibid.*, p. 232 (April 23, 1878).
53 *Ibid.*, pp. 223–224 (April 19, 1878); p. 232 (April 23, 1878); p. 234 (April 24, 1878).
54 In later letters, Chalmers mentions that he was "away in England from December 1877 to June 1878." *1879–1885 Exchequer Officials*, p. 413 (undated letter, *circa* November 17, 1882).
55 *1878 Personals (Vol. 1)*, p. 246 (June 9, 1878); *1878 Personals (Vol. 2)*, p. 9 (June 13, 1878).
56 *1878 Personals (Vol. 2)*, p. 111 (September 7, 1878); p. 126 (September 18, 1878); p. 170 (October 30, 1878).
57 *Ibid.*, p. 14 (June 18, 1878); p. 29 (June 27, 1878).
58 *1878 Personals (Vol. 1)*, p. 247 (June 9, 1878).
59 *1878 Personals (Vol. 2)*, p. 29 (June 27, 1868).
60 *Ibid.*, p. 13 (June 18, 1878); p. 25 (June 25, 1878); p. 29 (June 27, 1878).
61 *Ibid.*, p. 14 (June 18, 1878).
62 *Ibid.*, p. 177 (November 3, 1878).
63 *Ibid.*, p. 59 (July 20, 1878); pp. 70–72 (July 27, 1878); p. 75 (July 26, 1878).
64 *Ibid.*, p. 70 (July 27, 1878); p. 79 (July 29, 1878); p. 99 (August 28, 1878); p. 127 (September 18, 1878).
65 *Ibid.*, p. 173 (November 2, 1878).
66 *Ibid.*, p. 101 (August 28, 1878); p. 114 (September 7, 1878).
67 *Ibid.*, p. 104 (August 28, 1878).
68 *Ibid.*, pp. 90–95 (August 9, 1878); p. 119 (September 7, 1878).
69 Little could Chalmers know, of course, that Sutro's $3.5 million tunnel would be sold at a foreclosure sale in 1889.
70 *Ibid.*, p. 22 (June 25, 1878).
71 *Ibid.*, pp. 95–96 (August 9, 1878).

72 *Ibid.*, p. 87 (August 1, 1878).
73 *Ibid.*, p. 94 (August 9, 1878). Just a month earlier, Chalmers had written that it would take a cash offer of £10,000 to purchase the Leviathan. *Ibid.*, p. 32 (June 27, 1878).
74 *Ibid.*, pp. 77–78 (July 29, 1878).
75 *Ibid.*, p. 81 (August 25, 1878).
76 *Ibid.*, pp. 131–132 (September 29, 1878).
77 *Alpine Chronicle*, May 18, 1878.
78 *Alpine Chronicle*, August 31, 1878; September 14, 1878.
79 Dustman, K., *Silver Mountain City: Ghost of the Sierra* (Clairitage Press 2011), pp. 137–138.
80 *1878 Personals (Vol. 2)*, p. 126 (September 18, 1878).
81 *Ibid.*, p. 120 (September 7, 1878).
82 *Ibid.*, p. 116.
83 *Ibid.*, p. 128 (September 18, 1878).
84 *Ibid.*, p. 51 (July 5, 1878).
85 *Ibid.*, pp. 126–127 (September 18, 1878); p. 163 (October 28, 1878); p. 167 (October 28, 1878).
86 *Ibid.*, p. 51 (July 5, 1878).
87 *Ibid.*, p. 30 (June 27, 1878).
88 *Ibid.*, p. 155 (October 15, 1878); *see also Aberdeen Journal*, October 12, 1878, noting the Isabelle's subscription list was closed.
89 *Aberdeen Journal*, October 12, 1878.
90 *1878–1880 Isabelle Officials*, p. 41 (December 12, 1878); p. 54 (December 30, 1878).
91 *1878 Personals (Vol. 2)*, pp. 156–159 (October 23, 1878).
92 Chalmers had previously considered Ingersoll drills, but later indicated the National "beats them all, for many reasons." *Ibid.*, p. 154 (October 14, 1878); p. 176 (November 3, 1878); p. 184 (November 9, 1878); p. 198 (November 13, 1878).
93 *Ibid.*, p. 184 (November 9, 1878).
94 *Ibid.*, p. 185.
95 *Ibid.*, pp. 188–189 (November 9, 1878); p. 195 (November 11, 1878).
96 *Ibid.*, pp. 189–190 (November 9, 1878).
97 *Ibid.*, p. 207 (November 16, 1878) (*emphasis in original)*).
98 *Ibid.*, p. 191 (November 11, 1868).
99 *Ibid.*, p. 185 (November 9, 1878).
100 *Ibid.*, p. 176 (November 3, 1878); p. 210 (November 16, 1878).
101 *1878 Personals (Vol. 1)*, p. 202 (April 9, 1878); *1878 Personals (Vol. 2)*, p. 155 (October 14, 1878).
102 *1878 Personals (Vol. 2)*, pp. 194, 197 (November 11, 1878).
103 *Ibid.*, p. 176 (November 3, 1878).
104 *Ibid.*, p. 180.
105 *Ibid.*, pp. 195–196 (November 13, 1878) (*emphasis in original*).
106 *Ibid.*, p. 189 (November 9, 1878).
107 *Ibid.*, pp. 213–214 (November 18, 1878).
108 *Ibid.*, p. 213 (November 18, 1878).
109 *Ibid.*, pp. 216; 218 (November 18, 1878).
110 *Ibid.* p. 233 (December 3, 1878).
111 *Ibid.*, 1878 Ex'r Officials p. 249 (December 11, 1878).
112 *Ibid.*, pp. 223–224 (November 26, 1878); p. 227 (November 27, 1878).
113 *Ibid.*, pp. 229–230 (November 30, 1878).
114 *Ibid.*, pp. 229–30 (November 30, 1878).
115 *Ibid.*, p. 240 (December 9, 1878) (*emphasis in original*).
116 *Ibid.*, pp. 239–240.
117 *Ibid.*, pp. 236–237 (December 9, 1878).
118 *Ibid.*, p. 247 (December 10, 1878).
119 *Ibid.*, p. 249 (December 11, 1878).
120 *Ibid.*, pp. 245 (December 10, 1878).
121 *Ibid.*, p. 255 (undated list *circa* December 14, 1878).
122 *Ibid.*, p. 267 (December 19, 1878).
123 *Ibid.*, p. 265 (December 19, 1878).
124 *1878–1880 Isabelle Officials*, p. 30 (December 15, 1878).
125 *Ibid.*, p. 41 (December 21, 1878); p. 45 (December 25, 1878); *see also* p. 123 (February 4, 1879), reflecting a later £1,500 credit.
126 *Ibid.*, p. 53 (December 30, 1878).
127 *Ibid.*, pp. 31, 35 (December 15, 1878).

128 *1878 Personals (Vol. 2)*, p. 244 (December 8, 1878); p. 246 (December 10, 1878). Chalmers may have hired Hawkins as the surveyor this time because he still owed James Champion $75 for previous survey work; *see 1878 Personals (Vol. 2)*, p. 168 (October 28, 1878).
129 Before leaving San Francisco, Chalmers had requested that Arnot "have my Transit sent," as he planned to give it a "careful adjustment" before using it himself at the tunnel. *1878 Personals (Vol. 2)*, p. 221 (November 20, 1878).
130 *1878–1880 Isabelle Officials*, p. 30 (December 15, 1878); *Mono-Alpine Chronicle*, December 21, 1878; January 4, 1879.
131 *1878–1880 Isabelle Officials*, p. 38 (December 15, 1878).
132 *Ibid.*, p. 40 (December 20, 1878).
133 *Ibid.*, pp. 52–53 (December 30, 1878).
134 *1878–1880 Personals*, p. 1 (December 30, 1878) (transcription by Richard Race, copy at Alpine County Historical Society).
135 *1878–1880 Isabelle Officials*, p. 41 (December 21, 1878); p. 52 (December 30, 1878).
136 *Ibid.*, p. 52 (December 30, 1878).
137 *Ibid.*, p. 44 (December 25, 1878); p. 88 (January 25, 1879).
138 *Ibid.*, p. 31 (December 15, 1878).
139 *Ibid.*, p. 53 (December 30, 1878).
140 *Carson Valley News*, December 6, 1878, quoting a mining publication identified only as "*News Letter*, November 9."

Chapter 9

1 *1878–1880 Isabelle Officials*, p. 123 (February 4, 1879).
2 *Ibid.*, p. 58 (January 6, 1879); 67 (January 9, 1879).
3 *Ibid.*, p. 67.
4 *1878–1880 Personals*, p. 8 (February 18, 1879) (transcription by Richard Race, copy at Alpine County Historical Society).
5 *Ibid.*, pp. 15–16 (March 11, 1879).
6 *Ibid.*, p. 5 (February 12, 1879); p. 21 (May 7, 1879).
7 *Ibid.*, p. 12 (March 6, 1879).
8 *1878–1880 Isabelle Officials*, pp. 305–307 (July 23, 1879); p. 444 (December 3, 1879).
9 *1878–1880 Personals*, p. 19 (March 29, 1879).
10 *Ibid.*, p. 12 (March 6, 1879).
11 *Ibid.*, p. 24 (May 19, 1879).
12 *Ibid.*, p. 38 (September 10, 1879).
13 *Ibid.*, p. 14 (March 10, 1879); p. 17 (March 16, 1879).
14 *Ibid.*, p. 35 (June 25, 1879); *1879–1885 Exchequer Officials*, p. 30 (May 9, 1879).
15 *1879–1885 Exchequer Officials*, p. 47 (May 26, 1879).
16 *Ibid.*, p. 6 (April 30, 1879); *and see* unnumbered page immediately following (pasted in), mentioning "selling the I.X.L. wood over again." (April 30, 1879).
17 *1878–1880 Personals*, p. 23 (May 17, 1879); p. 27 (May 26, 1879); p. 34 (June 25, 1879); *1879–1885 Exchequer Officials*, p. 32 (May 9, 1879); p. 40 (May 14, 1879); p. 44 (May 14, 1879); and p. 51 (May 26, 1879).
18 *1878–1880 Personals*, p. 22 (May 17, 1879).
19 *1879–1885 Exchequer Officials*, p. 51–52 (May 26, 1879).
20 *Ibid.*, pp. 55–56 (June 6, 1879).
21 *Ibid.*, p. 40 (May 14, 1879); p. 44 (May 26, 1879); *1878–1880 Personals*, p. 23 (May 17, 1879).
22 *1879–1885 Exchequer Officials*, pp. 40–41 (May 14, 1879).
23 *1878–1880 Personals*, p. 22 (May 17, 1879).
24 *Ibid.*, p. 26 (May 26, 1879).
25 After considerable research, Chalmers purchased three National Drills instead of opting for the Ingersoll brand drill used by Sutro. He also used two National brand compressors. *1880–1881 Personals*, p. 371 (September 16, 1880).
26 *1878–1880 Isabelle Officials*, p. 212 (April 9, 1879); p. 214 (April 23, 1879); p. 225 (May 8, 1879); p. 258 (June 10, 1879); p. 263 (June 23, 1879); p. 269 (July 4, 1879); pp. 300–301 (July 17, 1879); p. 311 (July 27, 1879); p. 330 (August 9, 1879); p. 383 (September 27, 1879). Confirming it was the new compressor that was defective, *see Ibid.*, p. 225 (May 8, 1879).
27 *Ibid.*, p. 302 (July 21, 1879); p. 309 (July 28, 1879); p. 311 (July 27, 1879).
28 *Ibid.*, p. 313 (August 4, 1879); p. 365 (August 10, 1879).
29 *Ibid.*, p. 402 (October 13, 1879).

30 *Ibid.*, p. 299 (July 17, 1879).
31 *Ibid.*, pp. 281–283 (July 11, 1879).
32 *1879–1885 Exchequer Officials*, p. 102 (July 18, 1879).
33 *Ibid.*, p. 111 (July 21, 1879).
34 *Ibid.*, p. 85 (July 14, 1879).
35 *Ibid.*
36 *Ibid.*, p. 115 (July 21, 1879); p. 116 (July 22, 1879).
37 *Ibid.*, p. 98 (July 18, 1879).
38 *Ibid.*, p. 144 (August 25, 1879).
39 *Ibid.*, p. 87 (July 16, 1879); p. 101 (July 18, 1879).
40 *Ibid.*, p. 117 (July 23, 1879); p. 123 (July 27, 1879); p. 124 (July 28, 1879); p. 131 (August 18, 1879).
41 *Ibid.*, p. 124 (July 28, 1879); p. 131 (August 18, 1879).
42 *Ibid.*, p. 135 (August 18, 1879).
43 *Ibid.*, p. 142 (August 18, 1879); *1878–1880 Isabelle Officials*, p. 358 (September 8, 1879).
44 *Ibid.*, p. 303 (July 21, 1879); p. 308 (July 27, 1879); p. 317 (August 6, 1879); p. 323 (August 8, 1879); p. 333 (August 11, 1879); p. 338 (August 19, 1879); p. 343 (August 22, 1879); p. 351 (August 26, 1879); p. 361 (September 10, 1879); p. 382 (September 24, 1879).
45 *Ibid.*, p. 357 (September 8, 1879).
46 *Ibid.*, pp. 346–347 (August 24, 1879).
47 *Ibid.*, p. 338 (August 19, 1879); p. 341 (August 22, 1879); p. 343 (August 22, 1879). The Bandmann agent was a Mr. Birdsall.
48 *Ibid.*, p. 368 (September 10, 1879).
49 *Ibid.*, p. 261 (June 19, 1879); p. 277 (July 7, 1879); p. 285 (July 11, 1879); p. 287 (July 14, 1879).
50 *Ibid.*, p. 373 (September 12, 1879).
51 *Ibid.*, p. 406 (October 20, 1879).
52 *Ibid.*, p. 279 (July 11, 1879); p. 285 (July 11, 1879); pp. 289–292 (July 14, 1879); and pp. 490–491 (January 12, 1880).
53 *Ibid.*, p. 407 (October 16, 1879).
54 *Ibid.*, p. 323 (August 8, 1879).
55 *Ibid.*, p. 51 (30 December 1878).
56 *Ibid.*, p. 153 (February 19, 1879).
57 *Ibid.*, pp. 72–73 (January 17, 1879).
58 *Ibid.*, p. 227 (May 8, 1879); p. 247 (June 2, 1879).
59 *1878–1880 Personals*, p. 33 (June 18, 1879).
60 *1878–1880 Isabelle Officials*, p. 359 (September 8, 1879).
61 *Genoa Weekly Courier,* April 15, 1881. The letter was signed only "Markleeville." The Silver Mountain post office would finally close for good in 1883.
62 *Carson Valley News*, September 5, 1879. Arnot was commissioned by the California governor to the same post that November. *San Francisco Examiner*, November 12, 1879.
63 *1878–1880 Isabelle Officials*, p. 259 (June 9, 1879).
64 *Ibid.*, pp. 341–342 (August 22, 1879); *1879–1885 Exchequer Officials*, p. 143 (August 22, 1879).
65 *1878–1880 Isabelle Officials*, pp. 391, 393 (October 16, 1879).
66 *1878–1880 Personals*, p. 40 (September 22, 1879); p. 41 (September 24, 1879); p. 42 (September 29, 1879); p. 43 (October 13, 1879) (*emphasis added*).
67 *Ibid.*, p. 49 (October 30, 1879).
68 *Ibid.*, p. 44 (October 13, 1879).
69 *Ibid.*, p. 42 (September 29, 1879); p. 43 (October 16, 1879).
70 *Ibid.*, p. 42 (September 29, 1879); *see also* pp. 43 (October 16, 1879).
71 *Ibid.*, p. 44 (October 13, 1879).
72 *Ibid.*, p. 40 (September 22, 1879).
73 *1878–1880 Isabelle Officials*, p. 404 (October 16, 1879).
74 *1878–1880 Personals*, p. 44 (October 13, 1879).
75 *1878–1880 Isabelle Officials*, p. 438 (November 24, 1879); p. 452 (December 5, 1879); p. 467 (December 22, 1879).
76 *1878–1880 Personals*, p. 46 (October 20, 1879).
77 *Ibid.*, pp. 47–48 (October 31, 1879).
78 *1878–1880 Personals*, p. 51 (November 24, 1879); *see also* pp. 45–46 (October 16, 1879); p. 49 (October 30, 1879).
79 *Ibid.*, p. 50 (November 10, 1879).
80 *1878–1880 Isabelle Officials*, p. 407 (October 20, 1879); p. 418 (November 7, 1879).

81 The deeds were sent for recording in late December 1879. *1879–1885 Exchequer Officials*, p. 180–182 (December 26, 1879) and forwarded to London in January 1880. *Ibid.*, p. 186 (January 9, 1880).
82 *Ibid.*, pp. 157–158 (November 24, 1879).
83 *Ibid.*, p. 159 (November 24, 1879).
84 *1878–1880 Personals*, p. 55 (December 21, 1879).
85 *1878–1880 Isabelle Officials*, pp. 428–433 (November 10, 1879).
86 *Ibid.*, p. 455 (December 8, 1879).
87 *Ibid.*, p. 467 (December 22, 1879).
88 *Ibid.*, pp. 459–461 (December 10, 1879).
89 *1878–1880 Personals*, p. 54 (December 24, 1879); *see also 1878–1880 Isabelle Officials*, pp. 466–467 (December 22, 1879).
90 *1878–1880 Isabelle Officials*, p. 478 (December 29, 1879).
91 *1878–1880 Personals*, p. 53 (undated letter approximately December 8, 1879); pp. 54–56 (December 21, 1879).
92 *Ibid.*, p. 41 (September 24, 1879).
93 *Ibid.*, p. 52 (November 24, 1879).
94 *Ibid.*, p. 59 (December 29, 1879).
95 *1878–1880 Personals*, p. 57 (December 24, 1879); p. 56 (December 25, 1879).
96 *Ibid.*, p. 59 (December 26, 1879); p. 60 (December 29, 1879).
97 *1878–1880 Isabelle* Officials, p. 494 (January 12, 1880). Chalmers estimated the cost of the Isabelle at $26.78 or $27.98 per foot, depending on the factors included.
98 *Ibid.*, p. 461 (December 10, 1879); p. 465 (December 20, 1879).
99 *Ibid.*, p. 492 (January 13, 1880); p. 496 (January 12, 1880); pp. 509–510 (January 19, 1880).
100 *Ibid.*, pp. 509–510 (January 19, 1880); pp. 532–534 (February 2, 1880). The new blower and pipe alone "ran away with" $2,267.17 in materials and shipping. (*Ibid.*, p. 711 (5 May 1880).
101 *Ibid.*, p. 593 (March 8, 1880); p. 609 (March 15, 1880); *see similarly* letter to Syme, in *1878–1880 Personals*, p. 77 (March 6, 1880).
102 *1878–1880 Isabelle Officials*, p. 638 (March 24, 1880).
103 *Ibid.*, p. 621 (March 17, 1880).
104 *Ibid.*, pp. 668–669 (April 5, 1880).
105 George Deichmann was identified as a "Tin Smith" in the 1880 Federal Census for Alpine County (Township 2), enumerated 22–23 June 1880 as residing in the boarding house at Silver Creek.
106 *1878–1880 Isabelle* Officials, p. 604 (March 13, 1880).
107 *Ibid.*, pp. 583; 585 (March 4, 1880).
108 *Ibid.*, p. 593 (March 8, 1880).
109 *Ibid.*, p. 513 (January 27, 1880); p. 533 (February 2, 1880).
110 *1879–1885 Exchequer Officials*, p. 260 (March 9, 1880).
111 *1878–1880 Isabelle* Officials, p. 510 (January 19, 1880).
112 *1879–1885 Exchequer Officials*, p. 265 (March 13, 1880); p. 266 (March 17, 1880); p. 268 (March 18, 1880).
113 *Ibid.*, p. 280 (March 22, 1880).
114 *Ibid.*, pp. 243–244 (February 28, 1880); pp. 255–257 (March 10, 1880); p. 265 (March 13, 1880); p. 266 (March 17, 1880) (confirming that the taxes were paid and "both properties redeemed," having expended $4,918 to do so, the bulk of which ($3,270) was for arrears and penalties).
115 *Ibid.*, p. 249 (March 2, 1880).
116 *1878–1880 Personals*, p. 74 (March 4, 1880); p. 80 (March 8, 1880).
117 *Ibid.*, p. 65 (February 1, 1880); *see also* p. 69 (February 13, 1880).
118 *1880–1881 Personals*, p. 21 (March 23, 1880).
119 *1878–1880 Personals*, p. 75 (March 5, 1880); p. 76 (March 5, 1880).
120 *Ibid.*, p. 71 (February 27, 1880); *see similarly* p. 79 (March 8, 1880) (letter to Syme, asking "Is he [Beaumont] in with Cardeau to get me out?").
121 *Ibid.*, p. 80 (March 8, 1880).
122 *1879–1885 Exchequer Officials*, p. 208 (February 19, 1880); pp. 210–229 (February 21, 1880).
123 *Ibid.*, p. 226.
124 *1878–1880 Isabelle* Officials, pp. 599–607 (March 13, 1880).
125 *1879–1885 Exchequer Officials*, p. 214 (February 21, 1880).
126 *1878–1880 Isabelle* Officials, p. 664 (3 April 1880).
127 *Ibid.*, pp. 641–643 (March 26, 1880) (*emphasis added*).

128 *1878–1880 Personals*, p. 73 (March 2, 1880).
129 *1878–1880 Isabelle* Officials, p. 578 (March 3, 1880).
130 *1878–1880 Personals*, p. 77 (March 6, 1880).
131 *1880–1881 Personals*, pp. 3–4 (March 16, 1880); p. 21 (March 23, 1880).
132 *Ibid.*, p. 78 (13 April 1880).
133 *Ibid.*, p. 68 (April 9, 1880) (*inserted as a second page* 68).
134 *Ibid.*, pp. 50–52 (April 1, 1880).
135 *Ibid.*, p. 182 (June 11, 1880). Sometimes known as "Paris green," this emerald-colored dye was a highly toxic but beautiful coloring containing as much as fifty percent arsenic, known to produce weakness or even collapse in unsuspecting householders who inhaled traces of the poison.
136 *1878–1880 Isabelle* Officials, p. 649 (March 29, 1880).
137 *Ibid.*, pp. 695–696 (April 21, 1880) (offering one rig for $9,000); *1879–1885 Exchequer Officials*, p. 293 (5 April 1880) (describing two hoisting rigs, a Blake pump, safety cage and a large boiler for sale).
138 *1878–1880 Isabelle* Officials, p. 619 (March 18, 1880).
139 *Ibid.*, p. 634 (March 22, 1880).
140 *Ibid.*, Isabelle p. 691 (April 17, 1880).
141 *Ibid.*, p. 727 (May 7, 1880).
142 *Ibid.*, p. 704 (April 30, 1880); p. 730 (May 7, 1880).
143 *Ibid.*, p. 776 (June 30, 1880).
144 *Ibid.*, p. 658 (March 29, 1880); p. 670 (April 5, 1880).
145 *Ibid.*, p. 698 (April 26, 1880); *see also* p. 664 (April 3, 1880); p. 671 (April 5, 1880); p. 675 (April 6, 1880); p. 683 (April 12, 1880); p. 692 (April 19, 1880).
146 *Ibid.*, p. 698.
147 *Ibid.*, p. 786 (July 12, 1880); p. 790 (July 12, 1880); p. 796 (July 13, 1880).
148 *Ibid.*, p. 839 (July 31, 1880).
149 *Ibid.*, p. 683 (April 12, 1880); p. 692 (April 19, 1880); p. 696 (April 21, 1880); p. 697 (April 26, 1880).
150 *Ibid.*, p. 687 (April 17, 1880); *see also 1880–1881 Isabelle Officials*, p. 53 (September 13, 1880).
151 *1878–1880 Isabelle* Officials, p. 698 (April 26, 1880). Although signed by foreman Robert Andrews, the letter was in Chalmers's handwriting.
152 *Ibid.*, p. 711 (April 5, 1880).
153 *Ibid.*, p. 708 (May 3, 1880).
154 *Ibid.*, p. 716 (May 8, 1880); p. 739 (May 31, 1880).
155 *Ibid.*, p. 739 (May 31, 1880).
156 *Ibid.*, p. 700, (April 29, 1880).
157 *Ibid.*, pp. 755–756 (June 14, 1880).
158 *Ibid.*, pp. 761–3 (June 28, 1880).
159 *1879–1885 Exchequer Officials*, pp. 187–188 (January 9, 1880); *1880–1881 Personals*, p. 19 (March 22, 1880).
160 *1879–1885 Exchequer Officials*, p. 368 (February 5, 1881); *1880–1881 Personals*, p. 19 (March 22, 1880).
161 *1880–1881 Personals*, p. 19 (March 23, 1880).
162 *1879–1885 Exchequer Officials*, p. 301 (June 5, 1880).
163 *1880–1881 Personals*, p. 7 (March 16, 1880).
164 *1879–1885 Exchequer Officials*, p. 298 (May 8, 1880); p. 301 (June 5, 1880). Chalmers did eventually pay the cash installments. *Ibid.*, p. 323 (August 10, 1880); p. 332 (August 31, 1880).
165 *Ibid.*, p. 303 (July 12, 1880).
166 *Ibid.*, p. 323 (August 10, 1880) (indicating another £2,000 remittance was expected in August).
167 *1878–1880 Isabelle* Officials, p. 846 (August 16, 1880); *see also 1880–1882 Isabelle Officials*, pp. 2–18 (orders spanning August 25, 1880–August 30, 1880).
168 *1880–1882 Isabelle Officials*, p. 2 (August 25, 1880).
169 *1879–1885 Exchequer Officials*, p. 317 (July 30, 1880).
170 *1880–1882 Isabelle Officials*, p. 143 (November 16, 1880); *1879–1885 Exchequer Officials*, p. 353 (November 16, 1880); p. 354 (November 15, 1880). As part of this plan, Chalmers and a few trusted friends filed Applications to Purchase Timber Land. *Monitor Argus*, 6 September 1880 (original at Alpine County Museum). A legal appeal would be required to push it through, however. *1878–1880 Isabelle* Officials, p. 743 (June 1, 1880); p. 758 (June 24, 1880); p. 786 (July 12, 1880).
171 *Ibid.*, pp. 822–823 (July 27, 1880).

172 *1879–1885 Exchequer Officials*, p. 325 (August 10, 1880); *1878–1880 Isabelle* Officials, p. 857 (August 17, 1880).
173 *Ibid.*, p. 798 (July 13, 1880).
174 Williams, W., "Alpine County mansion was backdrop to history," *Modesto Bee* (September 2, 1979).
175 *1878–1880 Isabelle* Officials, p. 858 (August 18, 1880); *1880–1882 Isabelle Officials*, p. 2 (August 25, 1880).
176 *1878–1880 Isabelle* Officials, pp. 847–849 (August 16, 1880).
177 *1880–1881* Personals, p. 255 (July 16, 1880); p. 370 (September 16, 1880).
178 *Ibid.*, p. 415 (September 27, 1880); p. 432 (October 1, 1880); p. 574 (January 19, 1881); p. 754 (April 6, 1881).
179 *Ibid.*, pp. 233–234 (July 13, 1880).
180 *Ibid.*, p. 342 (August 30, 1880).
181 *Ibid.*, p. 289 (July 30, 1880).
182 *Ibid.*, p. 439 (October 4, 1880).
183 *Ibid.*, p. 287 (July 30, 1880).
184 *Ibid.*, p. 286 (July 30, 1880).
185 *Ibid.*, p. 308 (August 16, 1880).
186 *Ibid.*, pp. 337–339 (August 27, 1880).
187 *1880–1882 Isabelle* Officials, p. 73 (September 20, 1880).
188 *1879–1885 Exchequer Officials*, p. 335 (September 14, 1880).
189 *1880–1881 Isabelle Officials*, p. 18 (August 30, 1880).
190 *Ibid.*
191 *Ibid.*, p. 45 (September 10, 1880); p. 46 (September 13, 1880).
192 *Ibid.*, p. 94 (September 28, 1880); pp. 96–97 (October 4, 1880).
193 *Ibid.*, p. 116 (October 11, 1880).
194 *1880–1881 Personals*, p. 468 (October 25, 1880; *see also 1880–1882 Isabelle Officials*, p. 173 (December 16, 1880). The Diamond Drill had been ordered from L. Reynolds & Co. at 16 First Street, San Francisco.
195 *1880–1882 Isabelle Officials*, p. 98 (October 4, 1880); pp. 119–121 (October 25, 1880).
196 *Ibid.*, p. 127 (October 27, 1880); p. 140 (November 12, 1880); *see also 1880–1881 Personals*, pp. 464–471 (October 25, 1880); p. 525 (December 7, 1880).
197 Brewer, W. H., *Up and Down California* (1974 University of California Press), pp. 18; 26 (n.4); p. 352; *see also* portraits of Ashburner in photos after p. 168); *see also* Farquhar, F. P., *History of the Sierra Nevada* (1965 University of California Press), p. 130.
198 *1879–1885 Exchequer Officials*, p. 345 (November 12, 1880).
199 *Ibid.*, p. 346 (November 11, 1880).
200 *1880–1881 Personals*, p. 560 (January 13, 1881); p. 635 (February 16, 1881); p. 665 (February 19, 1881); *1880–1882 Isabelle Officials.*, pp. 224-225; 233 (February 10, 1881); p. 258 (February 22, 1881); p. 301 (April 14, 1881); p. 311 (April 20, 1881).
201 *1880–1882 Isabelle Officials*, p. 223 (January 29, 1881); p. 363 (October 21, 1881).
202 *1880–1881 Personals*, pp. 41–42 (March 26, 1880).
203 *Ibid.*, p. 352 (September 10, 1880).
204 *Ibid.*, p. 434 (October 4, 1880).
205 *Ibid.*, pp. 434–438 (October 4, 1880). Antoinette's symptoms suggest epilepsy or strychnine poisoning.
206 *1881–1884 Personals*, p. 258 (January 25, 1882).
207 *Ibid.*; *1880–1882 Isabelle Officials,* p. 163 (December 6, 1880); p. 173 (December 16, 1880).
208 *Ibid.*, pp. 202; 212 (January 22, 1881).
209 *Ibid.*, p. 162 (December 6, 1880).
210 *Ibid.*, p. 172 (December 15, 1880).
211 *Ibid.*, pp. 176–177 (December 18, 1880).
212 *Ibid.*, p. 183 (December 25, 1880); p. 184 (December 28, 1880).

Chapter 10

1 *1881–1884 Personals*, p. 258 (January 25, 1882).
2 *1880–1882 Isabelle Officials*, p. 274 (February 28, 1881).
3 *1879–1885 Exchequer Officials*, p. 373 (April 2, 1881).
4 *Ibid.*, p. 362 (January 11, 1881); p. 364 (January 19, 1881).
5 *1880–1882 Isabelle Officials*, pp. 224–226 (February 10, 1881).
6 *1879–1885 Exchequer Officials*, p. 366 (January 28, 1881).

7 *1880–1882 Isabelle Officials*, p. 223 (January 29, 1881); pp. 224 (February 10, 1881); p. 245 (February 18, 1881).
8 *Ibid.*, p. 311 (April 20, 1881).
9 *Ibid.*, pp. 265–271 (February 26, 1881).
10 *1880–1881 Personals*, pp. 727–728 (March 25, 1881); p. 794 (April 22, 1881).
11 *Ibid.*, p. 805 (April 27, 1881).
12 *1881–1884 Personals*, p. 5 (2 June 1881).
13 *1879–1885 Exchequer Officials*, p. 369 (March 7, 1881).
14 *1880–1881 Personals*, p. 819 (May 6, 1881).
15 *1881–1884 Personals*, p. 44 (July 26, 1881).
16 *Ibid.*, p. 48 (July 29, 1881); p. 86 (August 24, 1881).
17 *Ibid.*, p. 58 (August 12, 1881); p. 75 (August 14, 1881); pp. 79–80 (August 19, 1881); p. 84 (August 24, 1881).
18 *Ibid.*, p. 75 (August 14, 1881).
19 *Ibid.*, p. 29 (July 15, 1881).
20 *Ibid.*, p. 23 (July 13, 1881). The midwife Chalmers contacted was Mrs. Neddenriep, who helped deliver many babies in nearby Carson Valley, Nevada.
21 *Ibid.*, p. 56 (July 31, 1881).
22 *Ibid.*, p. 160 (October 24, 1881).
23 *Ibid.*, pp. 96_97 (September 11, 1881).
24 *Ibid.*, p. 159 (October 24, 1881).
25 *Ibid.*, p. 25 (July 13, 1881); pp. 45–47 (July 29, 1881); p. 53 (July 31, 1881); *1880–1882 Isabelle Officials*, pp. 349–350.
26 *1880–1882 Isabelle Officials*, p. 351.
27 *Alpine Chronicle*, December 5, 1868, *copy at* Alpine County Museum, Jones letter collection (#94.21.09g); *see also 1880–1882 Isabelle Officials*, p. 349 (August 15, 1881) (putting the Morning Star's tunnel at only 800 feet long when Chalmers took over).
28 *1881–1884 Personals*, p. 137 (October 2, 1881); pp. 66–67 (August 1, 1881).
29 *Ibid.*, p. 66–67 (August 14, 1881).
30 *Ibid.*, p. 137 (October 2, 1881).
31 *1880–1882 Isabelle Officials*, p. 364 (October 23, 1881); *see also 1881–1884 Personals*, p. 109 (September 14, 1881).
32 *1881–1884 Personals*, p. 112 (September 18, 1881).
33 *Ibid.*, p. 113.
34 *Ibid.*, p. 109 (September 14, 1881).
35 *1881–1884 Personals*, p. 117 (September 21, 1881).
36 *Ibid.*, p. 127 (September 26, 1881).
37 *Ibid.*, p. 117 (September 23, 1881).
38 *Ibid.*, p. 185 (November 27, 1881).
39 *Ibid.*, pp. 140–141 (October 2, 1881).
40 *Ibid.*, p. 141.
41 *Ibid.*
42 *Ibid.*, pp. 96–97 (September 11, 1881).
43 *Ibid.*, p. 157 (October 23, 1881).
44 *Ibid.*, p. 145 (October 16, 1881); p. 155 (October 21, 1881).
45 *1880–1882 Isabelle Officials*, pp. 364–365 (October 23, 1881).
46 *1881–1884 Personals*, p. 139 (October 2, 1881).
47 *Ibid.*, p. 137 (October 2, 1881).
48 *Ibid.*, p. 177 (November 18, 1881).
49 *1880–1882 Isabelle Officials*, p. 371 (October 26, 1881); p. 389 (November 14, 1881); *1881–1884 Personals*, p. 177 (November 18, 1881).
50 *1880–1882 Isabelle Officials*, pp. 399–400 (November 21, 1881); *1881–1884 Personals*, p. 176 (November 14, 1881).
51 *1881–1884 Personals*, p. 169 (November 14, 1881).
52 *Ibid.*, p. 305 (March 15, 1882).
53 Chalmers bestowed variations of his own first name on four of his children: Lewisa, his first, illegitimate daughter; first son Lewis, born in 1851; Jessica Louisa, born 1860; and, in 1881, son Lewis, with Antoinette.
54 *1881–1884 Personals*, p. 184 (November 27, 1881).
55 *Ibid.*, pp. 186–187 (*emphasis in original*).
56 *1880–1882 Isabelle Officials*, p. 464 (December 21, 1881); *see similarly* (*1881–1884 Personals*, p. 216 (December 21, 1881).

57 *1881–1884 Personals*, p. 202 (December 16, 1881); *1880–1882 Isabelle Officials*, p. 450 (December 12, 1881).
58 *1881–1884 Personals*, p. 215 (December 21, 1881).
59 *Ibid.*, p. 198 (December 12, 1881).
60 *Ibid.*, p. 203 (December 16, 1881); p. 204 (December 16, 1881).
61 *Mono-Alpine Chronicle*, January 4, 1879).
62 *1880–1881 Personals*, p. 96 (April 16, 1880).
63 *1881–1884 Personals*, p. 54 (July 31, 1881 Personals).
64 *Ibid.*, pp. 151–152 (October 21, 1881); *see also* p. 200 (December 14, 1881) (confirming the option was good until January 31, 1882).
65 *Ibid.*, p. 157 (October 23, 1881).
66 *Alpine Chronicle*, July 13, 1878; July 20, 1878; July 27, 1878).
67 *1881–1884 Personals*, p. 206 (December 16, 1881); p. 266 (February 5, 1882); *see also 1878–1880 Personals*, p. 77 (March 6, 1880); p. 78 (March 7, 1880) (*transcription by Richard Race, copy at Alpine County Historical Society*).
68 *1878–1880 Isabelle Officials*, p. 570 (February 25, 1880).
69 *1881–1884 Personals*, p. 207 (December 16, 1881).
70 *1881–1884 Personals*, p. 250 (January 25, 1882); correspondence dated March 24, 2011, from Chalmers family researcher Gordon Harland confirms that sister-in-law Eliza Chalmers died at Erlangen, Bavaria on November 4, 1881 (*copy in possession of the author*).
71 *1881–1884 Personals*, p. 98 (September 11, 1881).
72 *Ibid.*, p. 212 (December 18, 1881).
73 *1879–1885 Exchequer Officials*, p. 390 (December 26, 1881).
74 *1881–1884 Personals*, p. 218 (December 23, 1881); pp. 222–226 (December 23, 1881).
75 *Ibid.*, p. 192 (November 30, 1881).
76 *Ibid.*, p. 227 (December 23, 1881).
77 *Ibid.*, p. 254 (January 25, 1882).

Chapter 11

1 *1880–1882 Isabelle Officials*, pp. 507–509 (January 16, 1882).
2 *1881–1884 Personals*, p. 252 (January 25, 1882).
3 *1880–1882 Isabelle Officials*, pp. 510-511 (January 16, 1882).
4 *1881–1884 Personals*, p. 288 (February 26, 1882); *1879–1885 Exchequer Officials*, p. 405 (October 16, 1882).
5 *Ibid.*, pp. 262–263 (February 5, 1882).
6 *Ibid.*, p. 291 (February 26, 1882) (*emphasis in original*).
7 *Ibid.*, pp. 279, 281 (February 17, 1882).
8 *1880–1882 Isabelle Officials*, p. 646 (March 14, 1882); *1881–1884 Personals*, pp. 298–302 (March 12, 1882).
9 *1880–1882 Isabelle Officials*, p. 694 (April 6, 1882); [*1881–1884 Personals*, p. 341 (April 22, 1882)).
10 *1881–1884 Personals*, p. 342 (April 22, 1882); p. 310 (March 29, 1882) (confirming Chalmers's estimate of an "unlimited supply" of ore in the mine).
11 *1880–1882 Isabelle Officials*, p. 500 (January 12, 1882); p. 546 (January 25, 1882).
12 *Ibid.*, pp. 503–504 (January 16, 1882); p. 539 (January 25, 1882).
13 *1881–1884 Personals*, p. 283 (February 17, 1882); *see similarly* p. 276 (February 17, 1882).
14 *Ibid.*, p. 270 (February 10, 1882).
15 *Ibid.*, p. 272 (February 13, 1882).
16 *Ibid.*, p. 314 (April 10, 1882), p. 315 (April 7, 1882). Although the cause of Laura's death was not mentioned, she likely succumbed to tuberculosis like her older half-brother, based on Chalmers's mention of her 'weak lungs.'
17 *Ibid.*, pp. 346–347 (April 22, 1882).
18 *Ibid.*, p. 250 (January 25, 1882). John Chalmers had died about December, 1880. "[P]oor fellow, his lot lately has not been a pleasant one, but I never thought he would go before me," Chalmers wrote his son, Willie, adding: "Try and recover John's watch." *1880–1881 Personals*, pp. 553–554 (December 30, 1880).
19 *1881–1884 Personals*, p. 349 (April 22, 1882).
20 *Ibid.*, p. 325 (April 10, 1882). Johnny evidently wrote to his father again, as Chalmers replied two weeks later, "Perhaps the good fathers who so disinterestedly sent you to this Country may advance money enough to send you here to get rid of you," and provided travel directions. *Ibid.*, p. 354 (April 23, 1882).
21 *Ibid.*, pp. 248–250 (January 25, 1882).

22 *Ibid.*, p. 240 (January 9, 1882).
23 *Ibid.*, p. 236 (January 8, 1882).
24 *Ibid.*, p. 257 (January 25, 1882); pp. 287–288 (February 26, 1882); p. 304 (March 14, 1882).
25 *1880–1882 Isabelle Officials*, p. 525 (January 23, 1882); p. 528 (January 22, 1882); ten tons of cement copper would also be shipped to New York in September 1882. *1881–1884 Personals*, p. 482 (September 11, 1882). Cement copper was netting about 17 cents per pound. *Ibid.*, p. 313 (March 29, 1882).
26 *1881–1884 Personals*, p. 310 (March 29, 1882).
27 *Ibid.*, pp. 345–346 (April 22, 1882) (indicating he had made his own supply of calcium chloride); pp. 417–419 (August 29, 1882); p. 470 (September 8, 1882); pp. 480–485 (September 11, 1882); *1879–1885 Exchequer Officials*, p. 404 (October 16, 1882).
28 *1881–1884 Personals*, p. 295 (March 2, 1882).
29 *Ibid.*, p. 308 (March 29, 1882).
30 *Ibid.*, p. 348 (April 22, 1882).
31 *Ibid.*, p. 530 (October 8, 1882).
32 *Ibid.*, p. 534 (April 26, 1882).
33 *Ibid.*, pp. 421 *et seq.* (August 29, 1882); p. 501 (September 13, 1882); *1879–1885 Exchequer Officials*, p. 405 (October 16, 1882).
34 *1881–1884 Personals*, p. 502 (September 13, 1882).
35 *Ibid.*, p. 772 (July 16, 1883).
36 *Ibid.*, p. 521 (October 1, 1882).
37 *Ibid.*, p. 476 (September 11, 1882).
38 *Ibid.*, p. 489 (September 11, 1882).
39 *Ibid.*, p. 581 (December 18, 1882). Included in the $20,604 figure were thirty-nine bars of silver and gold bullion worth $18,172; the rest was presumably from copper.
40 *Ibid.*, p. 597 (December 24, 1882).
41 *Ibid.*, pp. 589–590 (December 24, 1882).
42 *Ibid.*, pp. 618–619 (January 10, 1883); p. 621 (January 15, 1883).
43 *Ibid.*, p. 646 (February 23, 1883).
44 *Ibid.*
45 *Ibid.*, p. 647.
46 *1879–1885 Exchequer Officials*, pp. 419–420 (January 5, 1883).
47 *1881–1884 Personals*, p. 651 (March 2, 1883).
48 *Ibid.*, p. 648 (February 23, 1883). Chalmers had put all his furniture into Antoinette's name to protect it from creditors, when they married. *Ibid.*, p. 664 (March 26, 1883).
49 *Ibid.*, p. 659 (March 15, 1883).
50 *Ibid.*, p. 662 (March 16, 1883).
51 *Ibid.*, p. 657 (March 9, 1883).
52 *Genoa Weekly Courier*, March 30, 1883.
53 *1881–1884 Personals*, p. 666 (March 24, 1883); pp. 676–677 (March 29, 1883).
54 *Ibid.*, p. 684 (April 11, 1883).
55 *Ibid.*, pp. 701–705 (May 23, 1883).
56 *Ibid.*, p. 707 (May 27, 1883).
57 *Ibid.*, p. 704 (May 23, 1883).
58 *Ibid.*, pp. 747–748 (June 22, 1883).
59 *Ibid.*, p. 711 (May 27, 1883).
60 *Ibid.*, pp. 693–694 (April 24, 1883); p. 727 (June 13, 1883) (confirming that a load of copper was to be credited to miner John Timblin's account).
61 *Ibid.*, p. 736 (June 11, 1883) (*emphasis in original*). Chalmers had concocted a similarly underhanded scheme to prevent attachment of the company's funds in Anglo California Bank, advising bank officials in San Francisco that checks for labor and supplies would be outstanding whenever bullion was being shipped. "In these circumstances, in the event of an attempted attachment, can you instruct your Cashier to answer 'no funds,'" he delicately suggested. *Ibid.*, pp. 621–622 (January 15, 1883).
62 *Ibid.*, pp. 757–760 (July 8, 1883).
63 *Ibid.*, pp. 766–767 (July 11, 1883) (*emphasis in original*).
64 *Ibid.*, p. 795 (August 27, 1883); p. 796 (August 27, 1883); p. 797 (September 2, 1883); pp. 801-802 (September 5, 1883).
65 *Ibid.*, p. 888 (January 12, 1884).
66 *Ibid.*, pp. 797–799 (September 2, 1883).
67 *Ibid.*, p. 889 (January 12, 1884).
68 *Ibid.*, pp. 803–804 (September 8, 1883).
69 *Ibid.*, p. 803 (September 8, 1883); p. 806 (September 10, 1883).

70 *Ibid.*, p. 808 (September 20, 1883); p. 809 (September 24, 1883); p. 814 (September 29, 1883).
71 *Ibid.*, p. 838 (November 4, 1883); pp. 844–845 (November 12, 1883).
72 *Ibid.*, p. 876 (December 23, 1883); p. 878 (December 24, 1883).
73 *Ibid.*, p. 826 (October 16, 1883); p. 834 (October 28, 1883).
74 *Ibid.*, p. 834 (October 28, 1883); p. 838 (November 4, 1883).
75 *Ibid.*, p. 863 (December 4, 1883).
76 *Ibid.*, pp. 861–865 (December 4, 1883); p. 878 (December 24, 1883); p. 879 (Chalmers's letter to Von Schmidt dated December 26, 1883).
77 *Ibid.*, p. 830 (October 21, 1883).
78 *Ibid.*, p. 833 (October 26, 1883).
79 *Ibid.*, p. 839 (November 5, 1883); p. 849 (November 20, 1883).
80 *Ibid.*, p. 891 (January 20, 1884). Chalmers promised Aaron he would pay out of his own pocket, if he could. *Ibid.*, p. 893 (January 19, 1884). That likely did not happen, either.
81 *Ibid.*, p. 813 (September 29, 1883); pp. 827–830 (October 21, 1883).
82 *Ibid.*, p. 846 (November 18, 1883).
83 *Ibid.*, p. 875 (December 23, 1883) (*emphasis in original*).
84 *Ibid.*, p. 889 (January 12, 1884).
85 *Ibid.*, p. 886 (January 13, 1884).
86 *Ibid.*, p. 887.
87 *Ibid.*, p. 892 (January 20, 1884).
88 *Ibid.*, p. 891 (January 20, 1884).
89 *Ibid.*, p. 881 (January 9, 1884); p. 891 (January 20, 1884).
90 *Ibid.*, p. 891 (January 20, 1884).
91 *Ibid.*, p. 894 (January 26, 1884).
92 *Ibid.*, pp. 895–897 (January 27, 1884).

Chapter 12

1 *1879–1885 Exchequer Officials*, pp. 427–428 (January 9, 1884). Mrs. Green had bought the property at the tax auction on February 26, 1883. *Ibid.*, p. 421 (April 11, 1883); p. 424 (December 19, 1883).
2 *Ibid.*, pp. 427–428.
3 Alpine County Records, Book "G," pp. 21–26 (May 12, 1884).
4 *Ibid.*, pp. 85–88 (March 7, 1885 conveyance, reciting Mrs. Green's earlier purchase).
5 *Aberdeen Weekly Journal*, August 2, 1884.
6 *1879–1885 Exchequer Officials*, p. 429 (November 18, 1884).
7 *Ibid.*, p. 431 (February 27, 1885).
8 Alpine County Records, Book "G," pp. 85–88 (March 7, 1885).
9 *Alpine Argus,* May 16, 1885.
10 *Ibid.*, June 13, 1885.
11 *Ibid.*, July 15, 1885.
12 *Ibid.*, September 19, 1885.
13 *Ibid.*, April 18, 1885. Annie Banks had married Alexander Chalmers in 1884.
14 *Ibid.*, June 6, 1885.
15 *1879–1885 Exchequer Officials*, p. 433 (June 29, 1885).
16 *Alpine Argus*, June 13, 1885.
17 Williams, Warren, "Alpine County Mansion Was Backdrop To History," *Modesto Bee*, September 2, 1979.
18 Alpine County Records, Book "G," p. 211 (August 31, 1885). The deed was signed in front of Peter Curtz and recorded at the request of Dominick Bari.
19 *Genoa Weekly Courier*, September 4, 1885; September 11, 1885.
20 *Aberdeen Journal*, July 5, 1886.
21 The Colorado No. 2 had passed into the hands of creditors in December 1880, after an execution sale was held in which even the tailings piles were sold. Alpine County Archives, Box 10, File 606.
22 *Alpine Argus*, September 25, 1886.
23 *Philipsburg Mail (Montana)*, September 8, 1887 (citing *Mining and Scientific Press*).
24 Alpine County Records, Book "G," pp. 555; 564; 601; 604 (November 10, 1897 to April 8, 1898).
25 Alpine County Records, Book "G," pp. 355–356 (December 12, 1888, not recorded until April 6, 1891).
26 *Daily Appeal (Carson City)*, July 23, 1889.

27 *Genoa Weekly Courier*, July 1, 1889.
28 *Ibid.*, October 18, 1889.
29 Receipt dated January 3, 1889, reflecting sale from N. D. Arnot to D. R. Hawkins (Holabird Western Americana auction listing). Arnot had acquired the Imperial mill site from John Scossa in 1887. Alpine County Records, Book "G," pp. 208–209 (May 5, 1887).
30 *Genoa Weekly Courier,* January 31, 1890, identifying John Weis as then-owner.
31 *Ibid.*, February 7, 1890; February 14, 1890.
32 Dustman, Karen, *Silver Mountain City: Ghost of the Sierra* (Clairitage Press 2011), p. 138.
33 Information from family genealogy researcher Gordon Harland (correspondence with the author, 2 July 2010).
34 *See* Lola Swartwood, "Silver Mountain City" (*The California Highway Patrolman*, April 1980), p. 72. Though this information was apparently based only on local rumor, it has a ring of truth to it; one could well imagine the aging Chalmers taking a clerical job to survive, with his fine penmanship, composition and bookkeeping skills.
35 *Genoa Courier*, March 9, 1891.
36 Alpine County Records, Book "G," pp. 355–356 (December 12, 1888).
37 The group of mining patents were for the Saugatuck, Fremont, and Buckeye No. 2 (*Ibid.*, Book "H", p. 181 (April 8, 1891)); the Accacia (Book "H," p. 186 (April 8, 1891)); and the Exchequer extensions (Book "H," p. 191 (June 17, 1891)).
38 *Genoa Weekly Courier*, June 5, 1891.
39 Warren Williams, *Modesto Bee*, *op. cit.*, citing Mrs. Eugenia Bruns's kindness to Antoinette.
40 *Genoa Weekly Courier*, June 5, 1891.
41 Letter from Lewis Chalmers to Dominick Bari (August 28, 1891), original at Alpine County Historical Society.
42 *Genoa Weekly Courier*, December 11, 1891.
43 *Ibid.*, March 25, 1892.
44 *Alpine County Records*, Book "G,", pp. 357–35 (May 2, 1891); *Ibid.*, pp. 444–445 (June 22, 1891).
45 *Ibid.*, Book "G," pp. 446–448 (January 16, 1892 purchase, recorded December 31, 1894).
46 *Liverpool Mercury*, August 7, 1893.
47 *Genoa Weekly Courier*, November 25, 1898. The same story was picked up in the *Carson Daily Appeal*, November 30, 1898).
48 *Genoa Weekly Courier*, August 18, 1899; *Carson Daily Appeal*, August 20, 1899.
49 Dustman, K., *Silver Mountain City: Ghost of the Sierra*, *op. cit.*, p. 138, n. 53; *see also* Wedertz, F., *Mono Diggings* (Community Printing and Publishing, 2001), p. 140.
50 Arnot apparently bid in only $825 of his $1,200 lien to acquire the Exchequer Mill property. *Carson Daily Appeal*, 13 February 1899. The Sheriff's Deed was recorded in Alpine County Records, Book "H," p. 197 (7 February 1900). On that same date, Arnot recorded the patents for the Saugatuck, Fremont, Buckeye No. 2, Accacia and Exchequer mines granted to Chalmers back in 1891.
51 1900 U.S. Census, Alpine County, California.
52 Alpine County Records, Book "H," pp. 292–293 (July 30, 1900).
53 Laura Chalmers remained enrolled through at least the spring of 1900, appearing on the school's honor roll that March. *Genoa Weekly Courier*, March 30, 1900. Now a teenager, Laura became friendly with the Curtz family of Monitor. In April 1901, Mrs. Curtz brought sixteen-year-old Laura along on a visit to Gardnerville. *Gardnerville Record*, April 4, 1901. Laura and Daisy Curtz were also sent off to school together for the September term, 1901. *Genoa Weekly Courier*, September 6, 1901.
54 Swartwood, L., "Silver Mountain City," *The California Highway Patrolman* (April 1980), p. 72.
55 Historical notes about the Chalmers family by Elizabeth Coyan (*circa* 1989, copy in possession of the author), whose parents had cared for Laura Chalmers at their ranch in Diamond Valley. Amazingly, Laura's suicide note has been preserved, and is in the Alpine County Historical Society collection.
56 *Record-Courier*, October 21, 1927.
57 *Record-Courier*, January 2, 1914.
58 Included in the household in 1901 were two of Ridder's daughters and their husbands, a twenty-seven-year-old clerk, and Chalmers. A 1902 postal directory for London shows Chalmers still identifying himself as a "mining engineer," with his business address remaining 55 and 56 Bishopsgate Street, London. Information from family genealogy researcher Gordon Harland (correspondence with the author, 2 July 2010).
59 Alex Chalmers died June 29, 1901. *Gardnerville Record*, July 5, 1901; *see also* Find-a-Grave Memorial #13175896.

60. *Genoa Weekly Courier*, October 11, 1889.
61. In 1894, Alex Chalmers sent a map to D. W. Virgin for use in a legal dispute involving Frey's Dam and Boyd's Lane; Alex was writing from White Plains, Churchill County, Nevada. Letter dated May 1, 1894, Nevada Historical Society, Virgin Papers Collection (MS #NC814/1/2).
62 *Gardnerville Record*, 31 July 1903. The article described the Isabella Tunnel as 9 × 11 feet "in the clear," wide enough for two mules to pass each other as they worked.
63 *Gardnerville Record*, August 7, 1903.
64 *Record-Courier*, May 20, 1904.
65 *Gardnerville Record*, August 7, 1903.
66 *Aberdeen Daily Journal*, March 5, 1904.
67 *Reno Gazette-Journal*, May 16, 1904.
68 *Record-Courier*, May 20, 1904.
69 *Ibid.*
70 Williams, W., *Modesto Bee*, *op. cit.*
71 Historical notes by Elizabeth Coyan, *op. cit.*; Obituary, *Oakland Tribune*, December 29, 1913.
72 1910 U.S. Census, Alpine County, California.
73 Historical notes by Elizabeth Coyan *op. cit.*
74 Brune, J., *Simple Balance: An Oral History* (Clairitage Press 2018), pp. 5–7.
75 Alpine County Records, Book "I," p. 438 (November 8, 1913).
76 Historical notes by Elizabeth Coyan *op. cit.*
77 *Oakland Tribune*, December 29, 1913.
78 *Record-Courier*, January 16, 1914.
79 Antoinette Chalmers's estate was probated in Alameda County, where she died. She had already decided to disinherit daughter Laura Carsten when she made out her will in November 1913. The total value of Antoinette's estate, including the $370 bank account and her right to receive $400 from the sale of her Alpine home, was just $908.25. Alameda County Superior Court, Probate Case No. 18477. (*FamilySearch.com*).
80 Coyan, E., Oral History, typescript at Alpine County Museum, p. 61; historical notes by Elizabeth Coyan *op. cit.*; *Reno Gazette-Journal*, March 11, 1947) (obituary for Henry Laughton, identifying his sister as Laura A. Ansty).
81 Conversation with Edie Blevins by the author, January 14, 2014.
82 *Record-Courier*, February 7, 1947; *Reno Gazette-Journal*, March 11, 1947; *Nevada State Journal*, March 12, 1947.
83 *Reno Evening Gazette*, March 13, 1947.
84 *FamilySearch*, Person ID #MBMH-MXM.
85 *FamilySearch*, Person ID #MBMW-4YF.
86 Information from Chalmers family genealogy researcher Gordon Harland (correspondence with the author, November 25, 2012).
87 Probate estate file of Dominick Bari, Alpine County Archives.
88 *Aberdeen Press and Journal*, February 11, 2012.
89 *Record-Courier*, October 21, 1927 (recollections of Harry Hawkins).
90 *1871–1873 Personals*, pp. 43 (1 May 1871); p. 50 (22 May 1871); p. 535 (21 March 1873).
91 As late as 1977, the total output of silver and gold from the entire Silver Mountain mining district was estimated at less than $300,000. Clark, W. B., *Mines and Mineral Resources of Alpine County* (California Division of Mines and Geology, 1977), p. 28; Dustman, K., *op. cit.*, p. 138, n. 54.
92 Comfort, H. G., *Camera Shots of Ghost Cities of California and Nevada* (Shafter Press 1941). Though no longer in Chalmers or Whitmore hands, the house continues to be privately owned.

Bibliography

Lewis Chalmers's letterbooks are a series of bound volumes, separated by year and purpose. His business correspondence was preserved in "Officials" for each of the three companies he managed (Imperial; Exchequer; and Isabelle). "Personals" volumes contained his family or personal correspondence, though details often overlapped with Officials of similar date, and occasionally a business letter would find its way into a "Personal" volume. The italicized titles below are my own reference system to identify each volume and are used in the Endnotes to this book.

Most, but not all, of these volumes are found in the collection of the Alpine County Historical Society (ACHS) in Markleeville, California. Where available, ACHS accession numbers are provided for those materials for ease of reference. Locations are listed for other materials.

As the reader will note, there are gaps in the chronology, indicating a few of Chalmers's volumes are missing. We can only hope that someday they will reappear.

Chalmers's elegant and somewhat idiosyncratic handwriting is difficult for modern readers to decipher. I have transcribed most (though not all) of the listed volumes, and plan to donate those transcriptions to ACHS. I hope this source list below and my transcriptions will help future researchers locate and access Chalmers's fascinating letters.

Personals

1868–1869 Personals—(spanning February 24, 1868–May 9, 1869). Volume at ACHS, #2005.001.001.

1869–1871 Personals—(spanning November 15, 1869–April 4, 1871). Volume at ACHS, #2005.001.006.

1871–1873 Personals—(spanning April 9, 1871–January 24, 1874). Volume at ACHS, #2005.001.007.

1876–1877 Personals—(spanning May 12, 1876–December 27, 1877). Volume at ACHS, #2005.001.002. (*Page references cited in the Endnotes refer to a transcription by Richard Race, which also may be found at ACHS.*)

1878 Personals (Vol. 1)—(spanning January 28, 1878–June 9, 1878). Volume at ACHS, #91.49.03.

1878 Personals (Vol. 2)—(spanning June 11, 1878–December 19, 1878). Volume at ACHS, #88.71.01.

1878–1880 Personals—(spanning December 30, 1878–March 16, 1880). Original volume location unknown; however, a transcription by Richard Race may be found at ACHS. (*Page references cited in the Endnotes refer to the transcription.*)

1880–1881 Personals—(spanning March 12, 1880–June 11, 1881). Volume at ACHS, #2005.001.003.

1881–1884 Personals—(spanning June 28, 1881–January 27, 1884). Volume at ACHS, #2005.001.005.

Imperial Officials

1867–1870 Imperial Officials—(spanning November 22, 1867–April 21, 1870). Volume at ACHS, #1988.071.02.

Exchequer Officials

1870–1873 Exchequer Officials—(spanning February 16, 1870–February 18, 1873). Volume at ACHS, #2009.002.001.
1873–1874 Exchequer Officials—(spanning March 3, 1873–September 10, 1874). Original volume is located at the Bancroft Library, San Francisco, California. ACHS holds a microfilm copy. The large-format scan, however, makes it difficult to read as constant side-to-side scrolling is required.
1879–1885 Exchequer Officials—(spanning April 28, 1879–June 29, 1885). Volume at ACHS, #88.71.03.

Isabelle Officials

1879–1880 Isabelle Officials—(spanning December 15, 1878–August 24, 1880). Volume at ACHS, #2005.001.004.
1880–1882 Isabelle Officials—(spanning August 25, 1880–May 8, 1882). Volume at ACHS, #1991.049.004.

Individual Letters

Two original Chalmers letters have survived that were not included in the letter books above:
1876 Letter from Chalmers to Board of Equalization regarding taxes—(July 21, 1876), original at Alpine County Historical Society, *Chalmers* materials folder.
1891 Letter from Chalmers to Dominick Bari—(August 28, 1891), original at Alpine County Historical Society, *Chalmers* materials folder.